REVISION NOTES IN PHYSICS

Book One

BOOKS BY M. NELKON

Published by Heinemann
ADVANCED LEVEL PRACTICAL PHYSICS (SI) *(with J. Ogborn)*
SCHOLARSHIP PHYSICS (SI)
OPTICS, SOUNDS AND WAVES (SI)
MECHANICS AND PROPERTIES OF MATTER (SI)
PRINCIPLES OF ATOMIC PHYSICS AND ELECTRONICS
REVISION NOTES IN PHYSICS (SI)
Book I, Mechanics, Electricity, Atomic Physics
Book II, Optics, Waves, Sound, Heat, Properties of Matter
GRADED EXERCISES IN PHYSICS (SI)
NEW TEST PAPERS IN PHYSICS (SI)
REVISION BOOK IN ORDINARY LEVEL PHYSICS (SI)
ELEMENTARY PHYSICS, Book I and II *(with A. F. Abbott)*
THE MATHEMATICS OF PHYSICS
(with J. H. Avery)
ELECTRONICS AND RADIO *(with H. I. Humphreys)*
SOLUTIONS TO ADVANCED LEVEL PHYSICS QUESTION (SI)
SOLUTIONS TO ORDINARY LEVEL PHYSICS QUESTIONS (SI)

Published by Chatto & Windus
PRINCIPLES OF PHYSICS (O-Level, SI)
EXERCISES IN ORDINARY LEVEL PHYSICS *(with SI units)*
C.S.E. PHYSICS
SI UNITS: AN INTRODUCTION FOR ADVANCED LEVEL

Published by Edward Arnold
ELECTRICITY (Advanced Level, SI)

Published by Blackie
HEAT (Advanced Level, SI)

BOOKS BY P. PARKER

Published by Heinemann
HEAT
ELECTRICITY AND ATOMIC PHYSICS

Published by Arnold
ELECTRONICS

REVISION NOTES IN PHYSICS

FOR ADVANCED LEVEL AND INTERMEDIATE STUDENTS

by

M. NELKON M.Sc(Lond.), F.Inst.P., A.K.C.

Formerly Head of the Science Department, William Ellis School, London

Book One

MECHANICS, ELECTRICITY ATOMIC PHYSICS

THIRD EDITION

(with SI units)

HEINEMANN EDUCATIONAL BOOKS LTD · LONDON

Heinemann Educational Books Ltd
LONDON EDINBURGH MELBOURNE TORONTO
AUCKLAND SINGAPORE JOHANNESBURG IBADAN
HONG KONG NAIROBI NEW DELHI
LUSAKA

ISBN 0 435 68640 2

THIRD EDITION © M. Nelkon 1971
First Published 1953
Reprinted 1954, 1955, 1956, 1957, 1958,
1959, 1960, 1962, 1964
Second Edition 1965
Reprinted 1966
Reprinted 1969 with amendments
Third Edition (with S I units) 1971
Reprinted 1974

Published by Heinemann Educational Books Ltd
48 Charles Street, London W1X 8AH
Filmset by Keyspools Ltd,
Golborne, Lancs
Printed in Great Britain by
Cox & Wyman Ltd, London, Fakenham and Reading

Contents

TO THE STUDENT

This book provides a quick revision of Mechanics, Electricity, and Atomic Physics. It does not replace any of your text-books on Physics, but supplements them. *Diagrams of the apparatus should be drawn when revising experiments.*

The following technique will be found useful when answering descriptive questions in Physics: (1) Draw a diagram first, and letter the various parts. (2) Then give the description or account, referring to the diagram by the letters. (3) If a measurement is required, *always give the formula for calculating the quantity concerned.* The formula should not be proved unless this is asked for in the question.

PREFACE TO THIRD EDITION

This edition covers the new Advanced level syllabus of the Examining Boards in Mechanics, Electricity and Atomic Physics. In the revised text, emphasis has been placed on B in electromagnetism and E in electric fields, in accordance with modern views, and the sections on motion of charges in fields, atomic physics, and motion of a rigid body have been expanded. SI units only are now used throughout the text, in accordance with the future policy of the Boards.

I am indebted to R. P. T. Hills, St. John's College, Cambridge, for his valuable assistance with this edition.

PREFACE TO FIRST EDITION

These Notes are intended to assist students taking the Advanced level or Intermediate examination in Physics. They are compiled from many years' experience of teaching and lecturing, and enumerate, compare, and contrast the principal points in Mechanics, Electricity, and Atomic Physics to this standard. The book is intended to supplement the existing text-books of Physics, and as it provides a summary of the work in a two-year Advanced level or Intermediate course, it is hoped that the Notes will prove useful to the second-year student for quick revision. Some degree of over-simplification has been unavoidable in condensing the subject-matter. Worked examples are included in illustration of the subject-matter.

The author is much indebted to Mr. A. F. Abbott, formerly Senior Science Master, Latymer Upper School, London, for reading the manuscript and for valuable suggestions; to Dr. J. Duffey, formerly of Watford Technical College, and Mr. C. F. Whorwell, William Ellis School, London, for reading the proofs; and to the following Examining Boards, for permission to reprint questions: University of London (L), Joint Matriculation Board (N.), Oxford and Cambridge Joint Board (O & C.), Cambridge Local Examinations Syndicate (C.), Joint Welsh Board (W.).

Mechanics

---

1. DYNAMICS. MOTION. ENERGY. GRAVITATION

LINEAR MOTION. WORK AND ENERGY

Summary of Fundamental Quantities.

QUANTITY	DEFINITION	VECTOR OR SCALAR	SI Unit
Velocity (v)	Rate of change of displacement.	Vector	**m s^{-1}**
Acceleration (a)	Rate of change of velocity.	Vector	**m s^{-2}**
Force (F)	Rate of change of momentum.	Vector	**newton (N)**
Mass (m)	Amount of matter	Scalar	**kg**
Momentum	Mass × velocity.	Vector	**kg m s^{-1}** or **N s**
Work	Force × distance.	Scalar	**joule (J)**
Energy	Capacity for doing work.	Scalar	
Potential energy	Energy by virtue of level or position.	Scalar	**J**
Kinetic energy	Energy by virtue of motion.	Scalar	**J**
Power	Rate of doing work.	Scalar	**watt (W)**

Note. (1) Acceleration due to gravity, $g = 9.8$ m s$^{-2} = 10$ m s^{-2} (approx.). Also, g = gravitational intensity = 9.8 N kg^{-1} = 10 N kg^{-1} (approx.).

(2) The *newton* is that force which gives a mass of 1 kg an acceleration of 1 metre per second^2.

(3) 1 kilogramme force (kgf) is the force due to gravity on a mass of 1 kg where g has the value 9.80665 m s^{-2}. 1 kilogramme weight (kg wt) varies all over the earth since g varies. Approximately, 1 kg wt = 9.8 N = 10 N.

(4) 1 joule = 1 newton × 1 metre. Fig. 1.1a.

Equations of linear motion. If an object undergoes a uniform acceleration a from an initial velocity u to a final velocity v in a time t, then, if s is the distance travelled:

$$v = u + at \quad \text{(i)} \qquad s = ut + \tfrac{1}{2}at^2 \quad \text{(ii)} \qquad v^2 = u^2 + 2as \quad \text{(iii)}$$

Note. (1) When an object falls under gravity, then $a = +g$, where g is 9.8 m s^{-2} (10 m s^{-2} approx.). If an object is thrown vertically upwards, then, for upward motion, $a = -g$; at the top of the flight, the velocity $v = $ o.

1

(2) Velocity, $v = \dfrac{ds}{dt}$. Acceleration, $a = \dfrac{dv}{dt} = \dfrac{d^2s}{dt^2} = v\dfrac{dv}{ds}$

Momentum. Force. Momentum $= mv$ . . . . (1)

Units: Momentum in kg m s^{-1} when m in kg and v in m s^{-1}

Force, $F =$ rate of change of momentum

$$= \frac{mv - mu}{t} \quad . . . \qquad (2)$$

$$\therefore F.t = mv - mu = \text{momentum change}$$

When m is constant, $F = m(v-u)/t = ma,$ . . . . (3)

where a is the acceleration produced.

Units: F in newtons (N) when m in kg and a in m s^{-2}

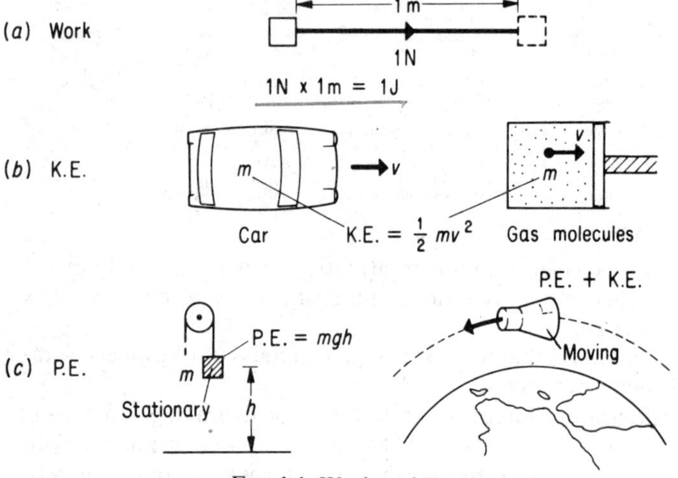

FIG. 1.1 Work and Energy.

Energy.   Kinetic energy, K.E. $= \frac{1}{2}mv^2$

Units: K.E. in J when m in kg, v in m s^{-1}

Kinetic energy may be *translational*, Fig. 1.1b, or *rotational*, as for a spinning wheel, or *vibrational*, as for a vibrating spring.

Potential energy, P.E. $= mgh,$

where h is the height above some level, such as sea-level, reckoned as "zero" P.E. Fig. 1.1c. A satellite in orbit has both K.E. (due to motion) and P.E. (due to height above the earth).

EXAMPLE

1. An object of mass 4 kg, moving with a velocity of 5 m s^{-1}, is brought to rest by a uniform retarding force F in (i) 2 s (ii) 2 m. Find F in each case.

(i) Momentum $= 4 \times 5 = 20$ N s

$$\therefore \quad F = \frac{\text{momentum change}}{\text{time}} = \frac{20}{2} = 10 \text{ N.}$$

(ii) Kinetic energy $= \frac{1}{2}mv^2 = \frac{1}{2} \times 4 \times 5^2 = 50$ J

$\therefore$ Work done against force $F \times s = F \times 2 = 50$ J, or $F = 25$ N

2. A projectile is fired from the ground at a speed of 400 m s^{-1} at an angle of 30° to the horizontal. Calculate (i) the greatest vertical height reached, (ii) the horizontal range. What is the minimum velocity needed for the same range? (Assume $g = 10$ m s^{-2})

(i) Vertical component of velocity $= 400 \sin 30° = 200$ m s^{-1}

From $v^2 = u^2 + 2as$ for vertical motion,

$0 = 200^2 - 2.10.s,$ or $s = 2000$ m $=$ maximum height.

(ii) From $s = ut + \frac{1}{2}at^2$ for vertical motion, time t when projectile reaches ground again is given by

$$0 = 400 \sin 30°. t - \frac{1}{2}. 10 . t^2, \quad \text{or} \quad t = 40 \text{ s}$$

$\therefore$ horizontal distance $=$ range $= 400 \cos 30° . t = 13860$ m

Minimum velocity. If α is the angle of inclination initially of the projectile velocity u to the horizontal, then $u \cos \alpha =$ horizontal velocity, $u \sin \alpha =$ vertical velocity. Hence time t for projectile to fall to ground is given by $0 = u \sin \alpha . t - \frac{1}{2}gt^2$ or $t = 2u \sin \alpha/g$. Thus range R is given by $R = u \cos \alpha . t = 2u^2 \sin \alpha \cos \alpha/g = u^2 \sin 2 \alpha/g$.
Maximum range is obtained when $2\alpha = 90°$. Thus maximum range $= u^2/g$.
Hence least velocity for range of 1386 m is given by $u^2/g = 1386$

$$\therefore \quad u = \sqrt{13860 \times 10} = 370 \text{ m s}^{-1}$$

Vectors and scalars. A *vector* quantity is one which has magnitude and direction; force, velocity, acceleration and momentum are vector quantities.

A *scalar* quantity is one which has magnitude only; mass, energy, power, temperature and volume are scalar quantities.

Resolved component. The resolved component (effective part) of a vector quantity F in a direction inclined at θ to F is given by

$$F \cos \theta.$$

The resolved component of F in a perpendicular direction is $F \cos (90 - \theta)$, or $F \sin \theta$. Fig. 1.2a.

Sum and differences of two vectors. The *resultant or sum, R*, of two vector quantities P, Q inclined at an angle θ to each other can be found by drawing P and then Q to scale, so that the two vectors form the successive sides of a triangle taken in order. The vector sum, $(\vec{P} + \vec{Q})$ is then represented by the third side.

Alternatively, drawn the "parallelogram of forces". The sum or resultant R is represented by the *diagonal* of the parallelogram. If P, Q are inclined at an angle θ to each other, then

$$R^2 = P^2 + Q^2 + 2PQ \cos \theta.$$

The *difference*, S, between the two vectors P, Q can be found by drawing OA to represent P, and AD to represent $-Q$; in the latter case, AD is drawn opposite in direction to Q. Then $S(\overrightarrow{P} - \overrightarrow{Q})$ is represented by OD in magnitude and direction.

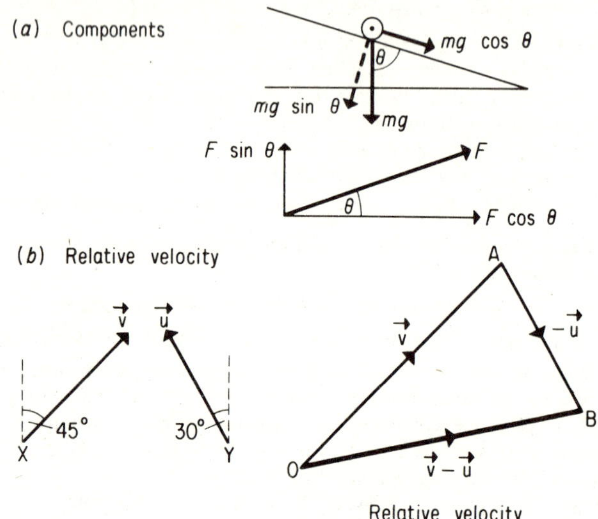

FIG. 1.2 Vectors.

Relative velocity.   This method can be applied to find the *relative velocity* between two objects moving in different directions. As an example, suppose we wish to find the velocity of a ship, X, moving at 20 km h^{-1} (v), N 45° E, relative to a ship Y moving at 15 km h^{-1} (u), N 30° W (Fig. 1.2b). We draw OA to represent 20 km h^{-1} completely ($\vec{v}$) and AB to represent -15 km h^{-1} ($-\vec{u}$) completely, i.e. AB is in the opposite direction to the velocity u. Then OB represents the velocity of X relative to Y in magnitude and direction, $\vec{v} - \vec{u}$. Similarly, BO represents the relative velocity of Y to X, $\vec{u} - \vec{v}$ (Fig. 1.2b).

Newton's laws of motion.   *Law I.* Every body continues in its state of rest or uniform motion in a straight line, unless impressed forces act on it.

Law II. The change of momentum per unit time is proportional to the impressed force, and takes place in the direction of the straight line along which the force acts.

Law III. Action and reaction are always equal and opposite.

Principle of Conservation of Linear Momentum. This states: "If no external forces act on a system of colliding objects, the total momentum of the objects remains constant."

The Conservation of Momentum Principle follows by combining Newton's law II with law III. Thus suppose a body A collides with a body B. If F is the force of A on B during collision, then, from law III, the force of B on A is equal and opposite to F. Thus if t is the time during collision, the quantity $F \times t$ is equal and opposite for A and B.

Now Ft = momentum change of A, and Ft = momentum change of B. Since the forces F are opposite, the *net* momentum change of A and B is zero. Hence the total momentum of A and B before collision = the total momentum of A and B after collision.

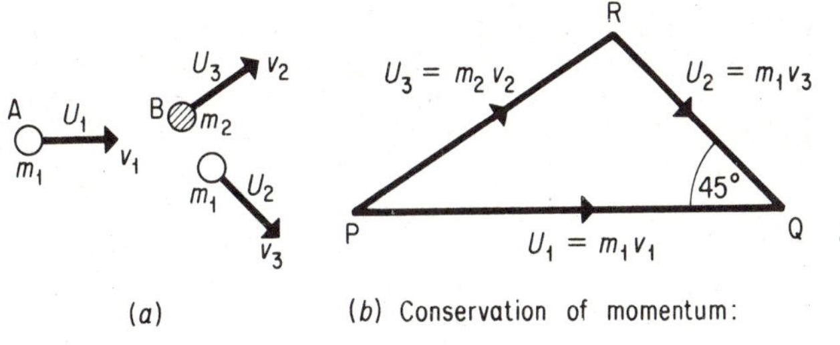

Elastic collision: $\frac{1}{2}m_1v_1^2 = \frac{1}{2}m_2v_2^2 + \frac{1}{2}m_1v_3^2$ (K.E. conserved)

Inelastic : $\frac{1}{2}m_1v_1^2 > \frac{1}{2}m_2v_2^2 + \frac{1}{2}m_1v_3^2$ (K.E. lost)

FIG. 1.3 Momentum and Energy in collisions.

Fig. 1.3 illustrates the collision of a moving object A with a stationary object B. The momenta (U_1, U_2, U_3) before and after the collision form a triangle, as shown. If no kinetic energy is lost in the collision, the latter is said to be *elastic*. Electrons colliding with gas atoms may make elastic collisions (p.158). If some kinetic energy is lost, the collision is said to be *inelastic*. In both elastic and inelastic collisions, the momentum is conserved.

If the mass of A = the mass of B = m say, and the collision is elastic, then $\frac{1}{2}mv_1^2 = \frac{1}{2}mv_2^2 + \frac{1}{2}mv_3^2$, so $v_1^2 = v_2^2 + v_3^2$. Hence triangle PQR has a right angle at R (Pythagoras). Thus A is deflected *perpendicular* to B after collision. Right-angle tracks can be seen after fast α-particles collide with helium atoms; showing that the masses are equal and the collisions are elastic.

Principle of Conservation of Energy. This states: "The total energy in a given system is constant, although energy may change from one form to another." Forms of energy are mechanical, heat, electrical, light, sound, and nuclear.

Problems in mechanics. When two objects collide, for example when a bullet or hammer strikes a target, or when two balls collide, the *conservation of linear momentum* must be used to find the new velocity. In writing down the total momentum the *direction* of the velocities must be carefully noted, since momentum is a vector quantity. It is wrong to use the conservation of mechanical energy principle when objects collide (unless the collision is elastic), because some mechanical energy is lost in the form of heat. On the other hand, the mechanical energy is conserved when an object moves from place to place without a collision, for example when the object falls from one height to another or swings in an arc from one position to another.

EXAMPLE

A sphere A(m_1) of mass 2 kg, initially moving with momentum $U_1 = 10{\cdot}0$ kg m s^{-1}, collides with a stationary sphere B of *equal* mass. A is deflected through $45°$ by the collision with a final momentum U_2 of $9{\cdot}0$ kg m s^{-1}. Find (*i*) the final momentum of B, (*ii*) the loss in energy of A.

(*i*) Fig. 1.3*a* shows the colliding spheres. From the conservation of linear momentum, the momentum of B after collision $= \vec{U_1} - \vec{U_2} = \vec{U_1} + (-\vec{U_2})$, where U_1 is the initial momentum of A and U_2 is the final momentum. In Fig. 1.3*b*, PQ is drawn to represent $\vec{U_1}$ and QR to represent $-\vec{U_2}$, to some convenient scale. Then PR represents completely the momentum U_3 of B after collision. Accurate measurement shows that PR $= 3{\cdot}7$ kg m s^{-1} at $60°$ to the initial direction of A.

(*ii*) Initial velocity of A $= 10{\cdot}0/2 = 5$ m s^{-1}. Final velocity $= 9{\cdot}0/2 = 4{\cdot}5$ m s^{-1}

$$\therefore \quad \text{loss in energy} = \tfrac{1}{2}mv_1{}^2 - \tfrac{1}{2}mv_2{}^2 = \tfrac{1}{2} \times 2 \times 5^2 - \tfrac{1}{2} \times 2 \times 4{\cdot}5^2 = 4{\cdot}8 \text{ J}$$

Dimensions. In terms of mass M, length L, time T:

Area	$= L^2$.	Volume	$= L^3$.
Velocity	$= L\,T^{-1}$.	Acceleration	$= L\,T^{-2}$.
Momentum	$= M\,L\,T^{-1}$.	Force	$= M\,L\,T^{-2}$.
Work, Energy	$= M\,L^2\,T^{-2}$.	Density	$= M\,L^{-3}$.

EXAMPLE

The flow of liquid in a tube becomes turbulent at some critical velocity v, which depends only on the viscosity η, the density ρ of the liquid and the radius r of the tube. Assuming η has dimensions $ML^{-1}T^{-1}$, find how v depends on η, ρ and r.

The dimensions of η are $ML^{-1}T^{-1}$; the dimensions of ρ are ML^{-3}; the dimension of r is L; the dimensions of v are LT^{-1}.

$$\therefore \quad LT^{-1} = (ML^{-1}T^{-1})^x (ML^{-3})^y (L)^z$$

Equating indices on both sides, we have

$$0 = x + y; \quad 1 = -x - 3y + z; \quad -1 = -x.$$

Solving

$$\therefore \quad x = 1, \quad y = -1, \quad z = -1$$

$$\therefore \quad v = \frac{k\eta}{\rho r}, \text{ where } k \text{ is a number.}$$

CIRCULAR MOTION

Angular velocity. ω = angular change per second = $d\theta/dt$: units are "radians per second" (rad s^{-1}). For an object making f revolutions per second, $\omega = 2\pi f$ in rad s^{-1}.

Velocity in circle. The velocity, v, is related to the angular velocity by $v = r\omega$, where r is the radius of the circle.

Acceleration in circle. When an object moves in a circle with uniform velocity, it has an acceleration *towards the centre* given by

$$\left(\text{acceleration} = \frac{v^2}{r} \text{ or } r\omega^2. \right)$$

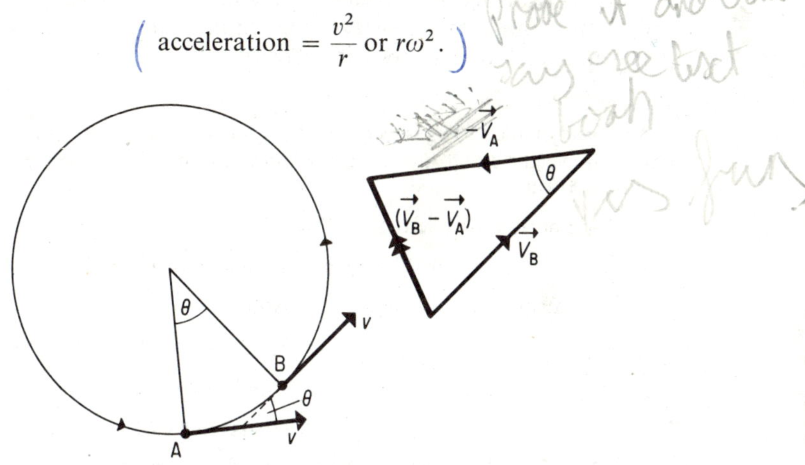

$$\text{Acceleration,} \quad a = \frac{\vec{v_B} - \vec{v_A}}{\delta t} = \frac{v^2}{r} \text{ or } \omega^2 r$$

FIG. 1.4 Acceleration in a circle.

Proof. Take two successive positions, **A**, **B**, of the object in a small time t (Fig. 1.4). Then velocity change = vector difference = $\vec{v}_B - \vec{v}_A = \vec{v}_B + (-\vec{v}_A)$. By drawing, the velocity change = the third side of an isosceles triangle whose two equal sides have magnitude $v = v.\theta$.

$$\therefore \quad \text{acceleration} = \frac{\text{velocity change}}{\text{time}} = \frac{v.\theta}{t} = v\omega = \frac{v^2}{r}.$$

Centripetal force = force towards centre = mv^2/r = $mr\omega^2$.

Banked track problems. For no side-slip at the wheels of a car going round a track, the only force on the car is the normal reaction R. Thus if θ is the angle of banking,

$$R \sin \theta = m \frac{v^2}{r} \text{ (towards centre)}$$

—7

$$R \cos \theta = mg \text{ (no vertical motion)}$$

$$\therefore \quad \tan \theta = \frac{v^2}{rg}.$$

Hence saucer-shaped track for racing, when v increases.

With side-slip, forces F act at wheels along the track and must be taken into account.

Centrifuges are machines which separate substances of varying density by whirling them round very fast. They are used to separate cream from milk and crystals from mother liquors, for example. The centripetal force varies with density.

EXAMPLE

An object of mass 10 kg is weighed at the poles and at the equator. Find the change in weight due to the earth's rotation, assuming it is spherical and of radius 6400 km.

$$\text{Period of rotation about axis, } T = 24 \text{ h} = 24 \times 3600 \text{ s}$$

$$\therefore \quad \text{angular velocity } \omega = 2\pi/T = 2\pi/(24 \times 3600) \text{ rad s}^{-1}$$

$$\therefore \quad \text{Centripetal force at equator} = mr\omega^2$$

$$= 10 \times 64 \times 10^5 \times (2\pi/24 \times 3600)^2$$

$$= 0.34 \text{ N (approx.)}$$

$$= \text{change in weight.}$$

At the poles, assuming $g = 9.8 \text{ m s}^{-2}$, weight $= mg = 10 \times 9.8 = 98$ N.

GRAVITATION

Kepler's laws. (1) The planets move in an ellipse round the sun, with the latter at one focus.

(2) The radius vector sweeps out equal areas in equal times.

(3) The square of the period is proportional to the cube of the average radius vector.

Newton's deduction. If m is the mass of a planet, ω its angular velocity, r the radius of a circle in which it moves uniformly, then force towards centre $= mr\omega^2$. *Assuming an inverse-square law,* the force $= Km/r^2$, where K is a constant.

$$\therefore \quad mr\omega^2 = mr\left(\frac{2\pi}{T}\right)^2 = \frac{Km}{r^2}$$

$$\therefore \quad T^2 = \frac{4\pi^2}{K} r^3, \quad \text{or} \quad T^2 \propto r^3.$$

This is Kepler's third law.

Newton's application to moon's motion. If the moon has a mass m, and the radius of its orbit round the earth is r, then force towards centre $= mr\omega^2$.

At the earth's surface the moon would be acted on by a force mg, and the distance from the mass of the earth is R, the radius of the earth's surface, imagining the mass of the earth concentrated at its centre. Since force $\propto 1/\text{distance}^2$,

$$\therefore\ \frac{mr\omega^2}{mg} = \frac{1/r^2}{1/R^2}$$

$$\therefore\ g = \frac{\omega^2 r^3}{R^2} = \frac{4\pi^2 r^3}{T^2 R^2}.$$

Substituting the values for T, r and R known in his day, Newton obtained an answer for g differing considerably from $9{\cdot}8\ \text{m s}^{-2}$. He heard of a new value of r some years later, and then found the calculated value of g to agree with its true magnitude.

Universal law of gravitation. Gravitational constant. The law of gravitation states that:

$$F = G\,\frac{m_1 m_2}{r^2},$$

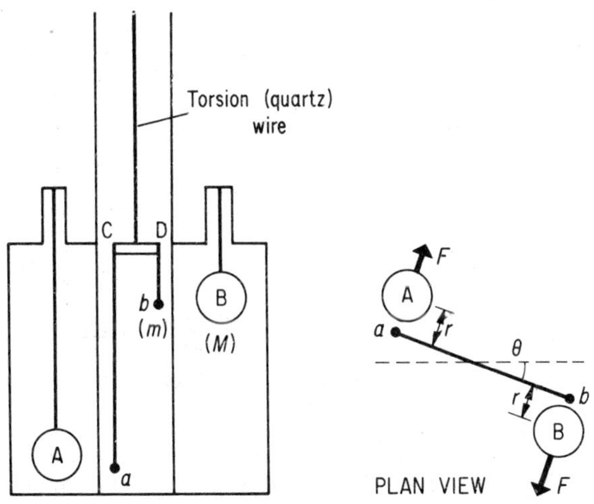

$$\text{Torque} = F.ab = G\frac{Mm}{r^2}\cdot ab = c\theta$$

$$\text{Period} = 2\pi\sqrt{\frac{I}{c}}\ \ (c\ \text{found})$$

Fig. 1.5 Boys' method for G.

where F is the force between two small masses m_1, m_2 which are distance r apart, and G is a constant called the *gravitational constant*. Note that G is a universal constant, whereas the value of g varies with the distance from the centre of the earth.

Method for G. *Boys' method.* 1895. Two large lead spheres A, B attracted two small gold balls a, b suspended by long and short quartz fibres from the ends of a highly polished bar CD, and the deflection θ observed (Fig. 1.5). Then, with the usual notation, $(GmM/d^2) \times CD = c\theta$. Oscillation of CD observed to find c, since $T = 2\pi\sqrt{I/c}$. Lamp and scale method for θ. High sensitivity of quartz fibres enabled accurate determination of θ. Small size of apparatus enabled it to be screened considerably from air convection currents. By experiment, $G = 6{\cdot}7 \times 10^{-11}$ N m^2 kg^{-2}.

Mass and density of earth. At the earth's surface, force on mass $m = mg = GMm/r^2$, where r is the earth's radius. Thus $M = gr^2/G$. Knowing g, G, r, then M is found to be about 6×10^{24} kg.

Assuming the earth to be a sphere, volume $4\pi r^3/3$,

$$\text{mean density } \rho \; = \; \frac{M}{V} \; = \; \frac{3g}{4\pi rG}. \; = \; 5500 \text{ kg m}^{-3}$$

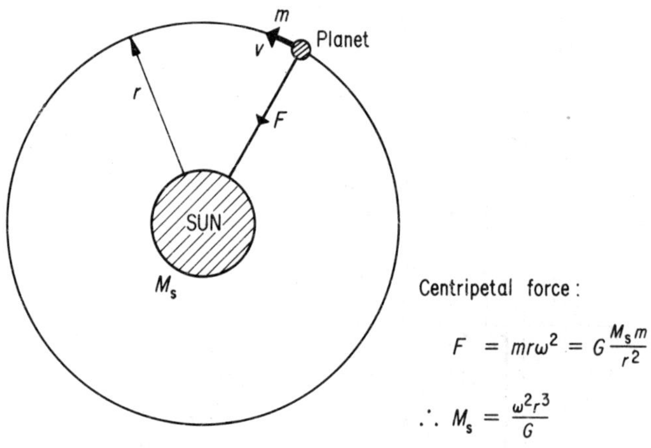

FIG. 1.6 Planetary motion.

If M_s is the *sun's* mass, then if m is the mass of a planet, $mr\omega^2 = GM_s m/r^2$. Fig. 1.6. Thus $M_s = \omega^2 r^3/G$ and can be found.

EXAMPLE

Obtain the value of g from the motion of the moon, assuming that its period of rotation round the earth is 27 days 8 hours and that the radius of its orbit is 60·1 times the radius of the earth, $6{\cdot}36 \times 10^6$ metre. If $G = 6{\cdot}7 \times 10^{-11}$ N m^2 kg^{-2}, find the mass of the earth.

At the earth's surface, $mg = \dfrac{GMm}{r^2}$, where r is the earth's radius and M is the mass of the earth .

$$\therefore \quad GM = gr^2 \qquad \ldots \ldots \ldots \ldots \ldots \ldots \qquad (i)$$

If m_1 is the mass of the moon, and R is the radius of its orbit round the earth,

$$\frac{GMm_1}{R^2} = m_1 R\omega^2$$

$$\therefore \quad GM = \omega^2 R^3 \qquad \ldots \ldots \ldots \ldots \ldots \ldots \qquad (ii)$$

From (i) and (ii),

$$gr^2 = \omega^2 R^3$$

$$\therefore \quad g = \omega^2 \times \frac{60 \cdot 1^3 r^3}{r^2} = \omega^2 \times 60 \cdot 1^3 \times r$$

$$= \frac{4\pi^2 \times 60 \cdot 1^3 \times 6 \cdot 36 \times 10^6}{(27\tfrac{1}{3} \times 24 \times 3600)^2}$$

$$= 9 \cdot 77 \text{ m s}^{-2}$$

From (i),
$$M = \frac{gr^2}{G} = \frac{9 \cdot 77 \times (6 \cdot 36 \times 10^6)^2}{6 \cdot 7 \times 10^{-11}}$$

$$= 5 \cdot 9 \times 10^{24} \text{ kg}$$

General. (1) *Inside earth.* If a point is distant r from the centre of the earth, the force of attraction on a mass m there $= GM_1 m/r^2$, where M_1 is the mass of a sphere *of radius r* of the earth. Now $M_1 = r^3 M/r_E{}^3$, where M is the mass of the earth, assuming the earth has uniform density for simplification

$$\therefore \text{ force of attraction} = \frac{GMmr}{r_E^3}, \quad \text{i.e. force} \propto r. \quad \text{Fig. 1.7}$$

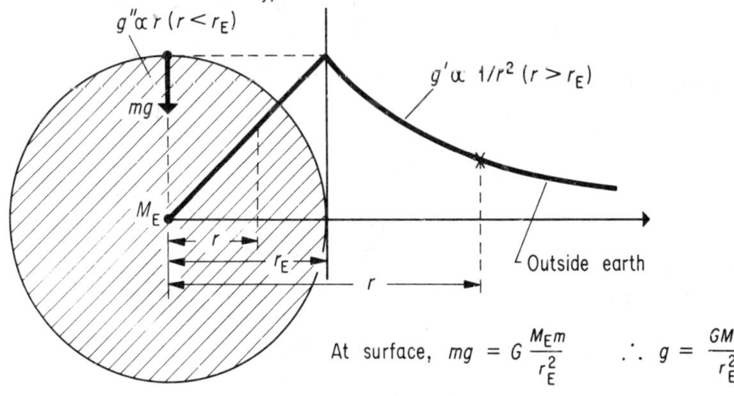

FIG. 1.7 Earth's gravitational field.

11

(2) *Outside earth.* Here we may assume that the mass of the earth is concentrated at the centre. Hence the acceleration g′ is proportional to $1/r^2$, where r is the distance from the centre. Fig. 1.7

Velocity of escape. The potential V at a point in the earth's gravitational field $= -GM/r$, where r is the distance from the centre. (The minus is due to the assumption that the potential at infinity is zero—compare the potential due to the nucleus of the atom). If a rocket of mass m is fired from the earth's surface X so that it just escapes from the gravitational influence of the earth, then loss in K.E. of rocket $= \frac{1}{2}mv^2 = m \times$ increase in potential between X and infinity.

$$\therefore \quad \tfrac{1}{2}mv^2 = m \times \frac{GM}{R},$$

where R is the radius of the earth.

$$\therefore \quad v = \sqrt{\frac{2GM}{R}} = \sqrt{2gR}, \quad \text{since } GM/R^2 = g.$$

Substituting $g = 9\cdot8$ m s^{-2}, $R = 6\cdot4 \times 10^6$ m, then $v = 11$ km s^{-1} (approx.) $=$ velocity of escape. Molecules in the atmosphere on the average have speeds less than velocity of escape. Little atmosphere round moon is due to smaller gravitational attraction.

Orbits. (1) If its velocity v is equal to the velocity of escape v_E, a satellite follows a parabolic path and never returns (Fig. 1.8). If its velocity v is greater than v_E, a satellite follows a hyperbolic path and never returns.

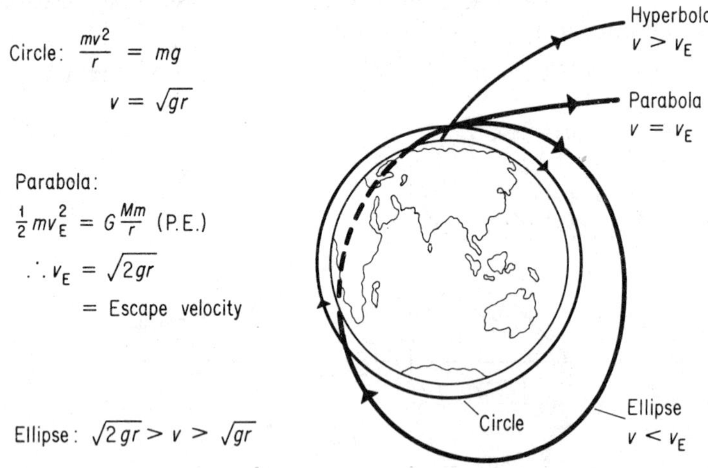

FIG. 1.8 Satellite orbits.

(2) When a satellite just circles round the earth, its velocity v_0 is given by

$$\frac{mv_0{}^2}{R} = \frac{GMm}{R^2},$$

or
$$v_0 = \sqrt{\frac{GM}{R}} = \sqrt{gR} = 8 \text{ km s}^{-1} \text{ (approx.)}$$

(3) If its velocity v is greater than 8 km s^{-1} and less than 11 km s^{-1} a satellite follows an elliptical path (Fig. 1.8).

2. OSCILLATIONS

SIMPLE HARMONIC MOTION

Definition of S.H.M.   This is the motion of a particle whose acceleration is proportional to its distance (y) from a fixed point and is always directed towards that point,

$$\text{i.e. acceleration} = -\omega^2 y.$$

S.H.M. relations. Simple harmonic motion can be regarded as the motion of the projection on a diameter of a particle moving round a circle with uniform speed v or $r\omega$. Fig. 2.1. Taking one position of the particle, and resolving the acceleration $r\omega^2$, towards the centre, along a vertical diameter we find that acceleration $= -\omega^2 y$.

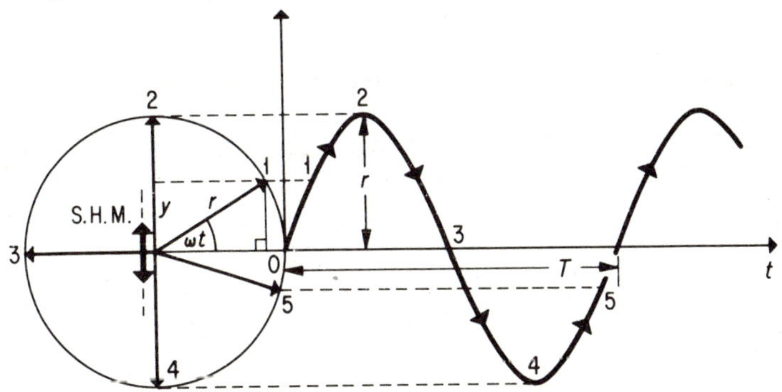

S.H.M. is the projection on a fixed diameter of circular motion:

$$a = -\omega^2 y$$

$$y = r \sin \omega t$$

$$v = \omega \sqrt{r^2 - y^2}$$

$$T = 2\pi/\omega$$

FIG. 2.1 Simple harmonic from circular motion.

The following formulæ hold for S.H.M.:—

Period, T, $= \dfrac{2\pi}{\omega}$.   *Acceleration* $= -\omega^2 y = -\omega^2 \times$ displacement.

Velocity at any instant, v, $= \omega\sqrt{r^2 - y^2}$; thus *maximum velocity* $= \omega r$, when $y = 0$ (centre of oscillation). At $y = r$ (end of oscillation), $v = 0$.

S.H.M. by calculus. Acceleration $= v\dfrac{dv}{dy} = -\omega^2 y$.

Integrating, $\qquad\qquad\qquad \therefore \quad \tfrac{1}{2}v^2 = -\tfrac{1}{2}\omega^2 y^2 + c$.

Since $v = 0$ when $y = r$, $\qquad\quad c = \tfrac{1}{2}\omega^2 r^2$.

$$\therefore \quad v^2 = \omega^2(r^2 - y^2)$$

$$\therefore \quad v = \omega\sqrt{r^2 - y^2}$$

$$\therefore \quad \frac{dy}{dt} = \omega\sqrt{r^2 - y^2}$$

$$\therefore \quad \omega t = \sin^{-1}\left(\frac{y}{r}\right) + c.$$

If $t = 0$ when $y = 0$, then $c = 0$.

$$\therefore \quad y = r\sin \omega t.$$

Simple Pendulum. For a small angle of swing (less than about 8°), the period T is given by

$$T = 2\pi\sqrt{\frac{l}{g}}.$$

This formula is proved by resolving the weight along the tangent at some instant, and using $F = ma$, i.e. $-mg\sin\theta = ma$ (see Fig. 2.2a). The minus indicates that the force is towards O but the displacement x is measured from O in the opposite direction.

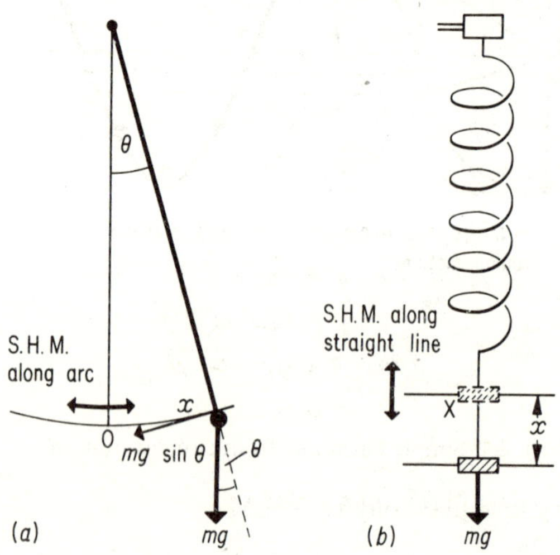

FIG. 2.2 S.H.M. in simple pendulum and spiral spring

14

Determination of g. Take 5 lengths l, time 50 oscillations for T, and plot l v. T^2. Then $g = 4\pi^2 \times$ gradient of straight line graph.

Pendulum suspended from a very high point. If distance of point from ground is h, and x is the height of the bob above the ground, $T = 2\pi\sqrt{(h-x)/g}$. If part of the suspension is cut off so that the height of the bob above the ground is x_1, then $T_1 = 2\pi\sqrt{(h-x_1)/g}$. By squaring both sides of the equations and eliminating h, g can be found.

Proportionality of mass and weight. If W is the weight of a pendulum bob of mass m, then, from force = mass × acceleration, we can show that the period is given by

$$T = 2\pi\sqrt{\frac{ml}{W}}.$$

$$\therefore \quad \frac{m}{W} = \frac{T^2}{4\pi^2 l}.$$

But T^2/l is a constant, by experiment. Hence $m/W = $ constant, or weight is proportional to mass.

Spiral spring or elastic thread. Provided Hooke's law is obeyed, the tension is proportional to the extension. If a weight X is put on and the extension is a, then $mg = ka$. Fig. 2.2b. When the weight X is pulled down a further distance and released, X vibrates *about* the original equilibrium position O. At a distance x below O,

$$-[k(a+x)-mg] = \text{net force} = m \times \text{accn.}$$

This simplifies to $\qquad \text{accn.} = -\frac{k}{m}x = -\omega^2 x,$

from which $\qquad T = \frac{2\pi}{\omega} = 2\pi\sqrt{\frac{m}{k}}$

$$\therefore \quad T^2/m = \text{constant.}$$

Determination of g by spiral spring. Taking into account the mass s of the spring, then

$$T = 2\pi\sqrt{\frac{m+\lambda s}{k}}$$

$$\therefore \quad \frac{k}{4\pi^2} T^2 = m+\lambda s.$$

Dynamic experiment. From a graph of T^2 v. m, $k/4\pi^2$ can be found (gradient required); hence k can be found.

Static experiment. By adding weights of mass M, and noting the extension a, a/M can be found from the slope of the graph. Now $Mg = ka$.

$$\therefore \quad g = \frac{a}{M} \times k.$$

Hence g can be found by combining the results of the static and dynamic experiment.

Free and damped oscillations Free and damped oscillations may be investigated by the apparatus shown in Fig. 2.3a. The amplitude

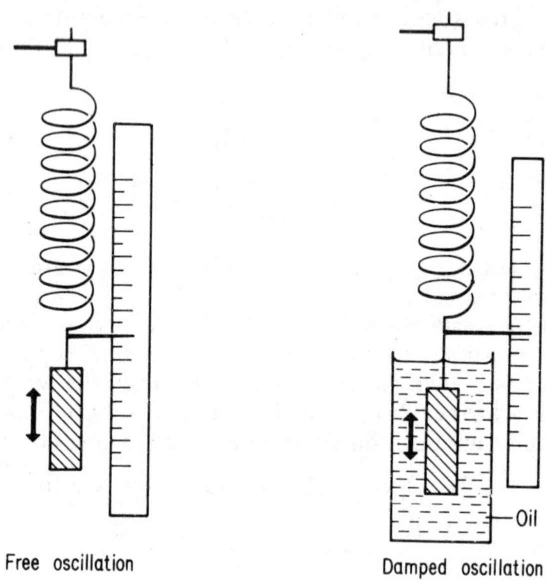

(a) Experiment

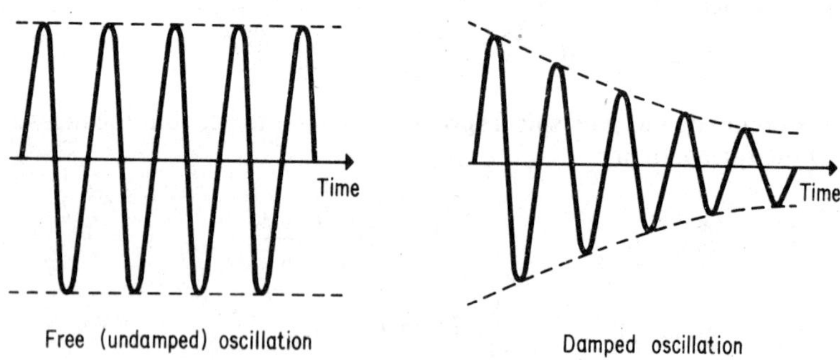

(b) Results

FIG. 2.3 Free and damped oscillations.

16

of vibration remains constant in free oscillations but diminishes exponentially under a resistive force proportional to the velocity. Fig. 2.3*b*.

Energy exchanges in S.H.M. (i) Centre of oscillation: K.E. = maximum, P.E. = 0. (ii) End of oscillation: K.E. = 0, P.E. = maximum. Interchange of P.E. and K.E. always occurs during S.H.M.
 Total energy = maximum K.E. = $\frac{1}{2} m a^2 \omega^2$.

 Examples: (1) *Mechanical.* Fig. 2.4*a*. Oscillation of mass *m* attached to a spring on a smooth horizontal table—here the kinetic energy, K.E., of the mass is
(a) zero when the spring is fully extended (P.E. of spring a maximum)·
(b) a maximum when the spring extension is zero (P.E. of spring = 0).

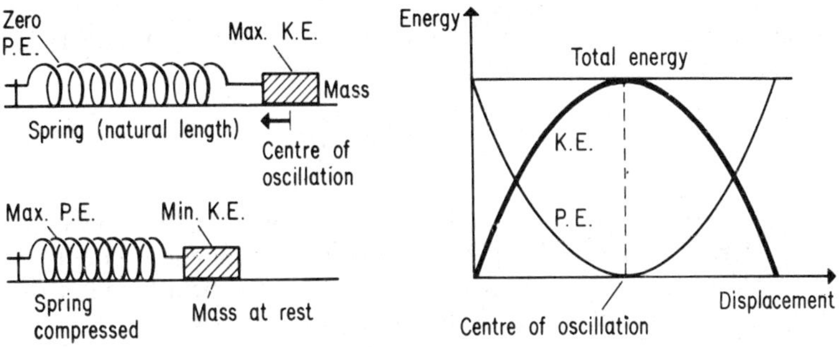

(*a*) Mechanical oscillations

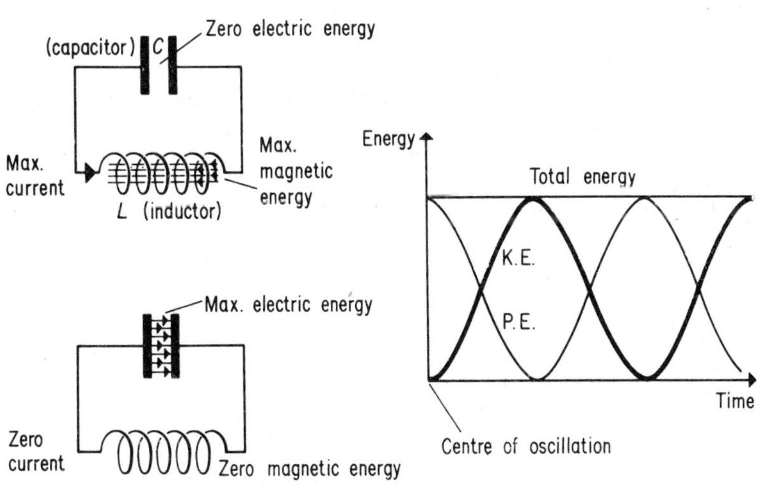

(*b*) Electrical oscillations.

FIG. 2.4 Energy exchanges in S.H.M.

17

The *total* energy of the system (spring+mass) = the maximum P.E. of the spring = $\frac{1}{2}ka^2$, where a is the amplitude of motion of the mass and $F = kx$ for the spring. Thus if v is the velocity of the mass at some instant and x is then the spring extension,

$$\text{total energy} = \tfrac{1}{2}mv^2 + \tfrac{1}{2}kx^2 = \tfrac{1}{2}ka^2$$

$$\therefore \quad v^2 = \frac{k}{m}(a^2 - x^2)$$

$$\therefore \quad v = \sqrt{\frac{k}{m}}\sqrt{a^2 - x^2}.$$

Comparing this expression for v with $v = \sqrt{\omega(a^2 - y^2)}$ for S.H.M. on p. 14, it follows that the period T is given by

$$T = \frac{2\pi}{\omega} = 2\pi\sqrt{\frac{m}{k}}.$$

(2) *Electrical.* Fig. 2.4*b*. Coil (L) – capacitor (C) in electrical circuit. Here the energy changes continually from magnetic energy in the field of the coil to electrostatic energy in the field between the capacitor plates. When the former energy is a maximum (i.e. maximum current flowing), the electrostatic energy is zero (no voltage across plates at this instant.).

EXAMPLE

A particle of mass 20 g moves with S.H.M. of frequency 10 Hz and amplitude 10 cm. Calculate (*a*) the maximum velocity of the particle, (*b*) its velocity when the displacement from the equilibrium position is 6 cm, (*c*) its potential and kinetic energy in case (*b*).

The velocity is given generally, see above, by

$$v = \omega\sqrt{a^2 - x^2}.$$

(*a*) $\therefore$ maximum velocity $= \omega a$ (when $x = 0$) $= (2\pi \times 10) \times 10 = 628$ cm, since $\omega = 2\pi f$.

(*b*) $v = \omega\sqrt{a^2 - x^2} = 2\pi \times 10 \times \sqrt{10^2 - 6^2} = 500\,\text{cm s}^{-1}$ (approx.) $= 5\,\text{m s}^{-1}$.

(*c*) K.E. $= \frac{1}{2}mv^2 = \frac{1}{2} \times (20 \times 10^{-3}) \times 5^2$ J $= 0.25$ J.

Total energy in vibration = maximum K.E. $= \frac{1}{2} \times (20 \times 10^{-3}) \times 6.28^2$ J

$$= 0.39 \text{ J (approx.)}$$

$$\therefore \quad \text{P.E.} = \text{total} - \text{K.E.} = 0.39 - 0.25 = 0.14 \text{ J}$$

3. MOTION OF RIGID BODIES. STATICS. FLUIDS

MOTION OF RIGID BODIES

Moment of inertia (I). *Definition,* $I = \Sigma mr^2 = Mk^2$, *where* k is the radius of gyration.

Uniform rod about axis through (i) *middle perpendicular to length* $l = Ml^2/12$ (ii) *one end perpendicular to length* $= Ml^2/3$.

Circular disc about axis through centre perpendicular to plane, $I = Ma^2/2$ (a = *radius*). *Cylinder about axis of symmetry,* $I = Ma^2/2$. *Sphere about diameter.* $I = \frac{2}{5}Ma^2$.

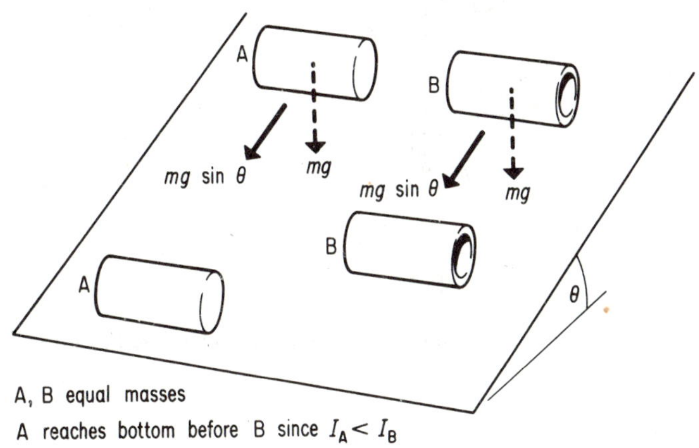

FIG. 3.1 Effect of moment of inertia.

Fig. 3.1 shows two cylinders of the same weight, *mg*, rolling down an inclined plane from the same height. A is solid, B is hollow. A has the smaller moment of inertia, since its mass is distributed on the average nearer to the axis than that of B. Hence A reaches the bottom first (See *Conservation of energy*, p. 21).

Theorem of parallel axes. $I_O = I_G + Mh^2$,

where I_G is the moment of inertia about the centre of gravity G, and h is the distance of G from the axis O.

The moment of inertia, I, about an axis on the circumference of a hoop of radius a perpendicular to its plane is thus given by

$$I = I_G + Mh^2 = Ma^2 + Ma^2 = 2Ma^2,$$

since $I_G = Ma^2$ and $h = a$.

Theorem of perpendicular axes. $I_z = I_x + I_y$, where x, y, z are mutually perpendicular axes.

Couple or Torque on rotating object $= I \times$ ang. accn. $= I d\omega/dt$. In Fig. 3.2, $mg.r = I \times$ angular acceleration of flywheel.

Work done by torque $=$ torque $\times$ angle of rotation in radians.

Kinetic energy of rotation about axis $= \frac{1}{2}I\omega^2$. Fig. 3.2.

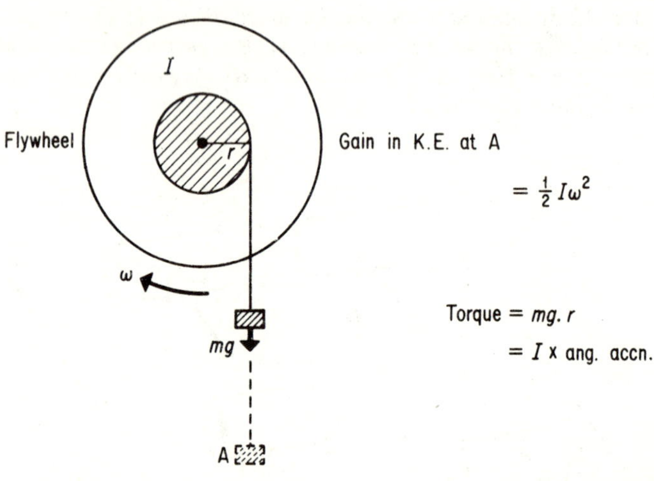

FIG. 3.2 Torque (Couple) on rotating object and K.E. gain

EXAMPLE

A flywheel of mass 10 kg and radius 20 cm is mounted on a central axle of radius 1 cm. Find (*i*) the angular acceleration produced by a constant force of 2 kgf tangential to the axle, (*ii*) the kinetic energy gained in 10 complete revs.

(*i*) Moment of inertia of flywheel, $I = \dfrac{Ma^2}{2} = 10 \times \dfrac{0 \cdot 2^2}{2} = 0 \cdot 2$ kg m^2

Force $= 2$ kgf $= 20$ N approx., hence torque $C = 20 \times 0 \cdot 01 = 0 \cdot 2$ newton metre. From $C = I \times$ angular acceleration,

$$\text{angular acceleration} = \frac{C}{I} = \frac{0 \cdot 2}{0 \cdot 2} = 1 \text{ rad s}^{-2}$$

(*ii*) K.E. gained $=$ work done $=$ torque $\times$ angle of rotation in radians

$$= 0 \cdot 2 \times 2\pi \cdot 10 = 12 \cdot 6 \text{ J}$$

Kinetic energy of rotation and translation $= \frac{1}{2}Mv^2 + \frac{1}{2}I\omega^2$, where M is the mass of the object, v its velocity, I is the moment of inertia about the c.g., and ω its angular velocity *about the c.g.*

Conservation of energy. When objects collide, some mechanical energy is lost in the form of heat. When an object does not collide, its mechanical energy is constant. Suppose a cylinder rolls a distance s down a plane from rest. Then, since gain of kinetic energy = loss of potential energy,

$$\tfrac{1}{2}Mv^2 + \tfrac{1}{2}I\omega^2 = \text{K.E.} = Mgs \sin \alpha \text{ (P.E. loss)},$$

where α is the angle of inclination of the plane.

Note that v, the velocity of the centre of gravity, $= a\omega$, where a is the radius of the cylinder. We can therefore substitute $a\omega$ for v (or v/a for ω) in the energy equation. The acceleration a down the plane would be found by applying the relation $v^2 = 2as$. The result shows that $a \propto 1/I$. Thus a cylinder of small I rolls fast down a plane (see Fig. 3.1, p. 19).

Angular momentum. Conservation of angular momentum.

Angular momentum = moment of momentum
$$= I\omega.\ \text{Fig. 3.3}a.$$

(a) Angular momentum about $0 = I\omega$

(b) Planetary orbit

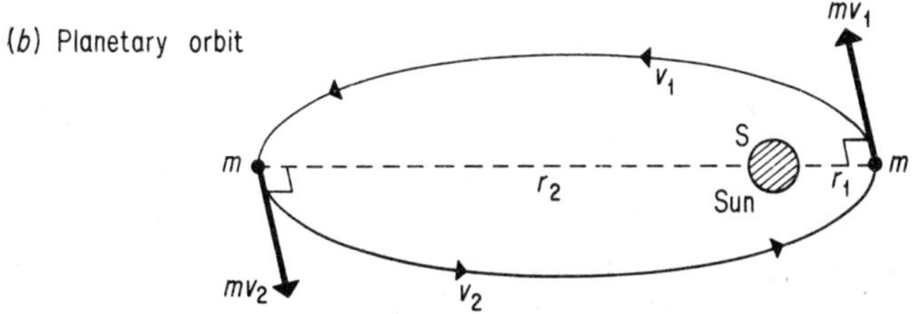

Conservation of angular momentum about S: $\therefore mv_1r_1 = mv_2r_2$

$$\therefore v_1 > v_2$$

Fig. 3.3 Angular momentum and conservation

The conservation of angular momentum states that the angular momentum about an axis of a given rotating body or system of bodies is a constant if no external forces act. In planetary motion round the sun S, the external force of attraction passes through S and hence has no torque about S. Thus the angular momentum of a planet about S is constant. Fig. 3.3b. If a ring is placed concentrically and gently on a spinning disc, for example, the conservation of angular momentum can be used to find the reduced speed of the disc.

Period of oscillation. For a rigid body of mass m oscillating about a fixed horizontal axis O,

$$I \times \text{ang. accn.} = \text{torque} = mgh.\theta,$$

where h is the distance OG (G is C.G.), θ is the *small* angle made by OG with the vertical. Thus

$$\text{period } T = \frac{2\pi}{\sqrt{mgh/I}} = \left(2\pi\sqrt{\frac{I}{mgh}}\right)$$

Compound pendulum. $I = I_G + mh^2 = mk^2 + mh^2$.

$$\therefore \quad T = 2\pi\sqrt{\frac{k^2 + h^2}{hg}} = 2\pi\sqrt{\frac{l}{g}},$$

where $(k^2 + h^2)/h = l$.

$$\therefore \quad h^2 - hl + k^2 = 0$$
$$\therefore \quad h_1 + h_2 = l, \quad \text{and} \quad h_1 h_2 = k^2.$$

By measuring the period T at different distances h from the centre of gravity, a graph of T v. h is drawn. The two values of h giving the *same* period, h_1 and h_2, are then found from the graph, and g is calculated from

$$T = 2\pi\sqrt{\frac{l}{g}} = 2\pi\sqrt{\frac{h_1 + h_2}{g}},$$

i.e.

$$g = \frac{4\pi^2(h_1 + h_2)}{T^2}.$$

Note that $k^2 = h_1 h_2$, so that at the *minimum* value of T, where $h_1 = h_2$, $h_1^2 = k^2$ or $h_1 = k$. The moment of inertia about the C.G. $=$ mass of pendulum $\times k^2$ and can hence be found.

Determination of moments of inertia. Flat plate. (i) Find the period T of *torsional* horizontal oscillations when the plate, moment of inertia I, is suspended by a wire; (ii) then find the new period T_1 with an annulus (ring) of moment of inertia I_R on the plate. Then, since $T = 2\pi\sqrt{I/c}$ and $T_1 = 2\pi\sqrt{(I + I_R)/c}$,

$$\frac{T}{T_1} = \sqrt{\frac{I}{I + I_R}},$$

or

$$I = \frac{T^2}{T_1^2 - T^2} \cdot I_R. \quad \text{(See Fig. 3.4}a)$$

Flywheel about horizontal axle. Attach one end of string to a pin on the axle, wind the string round the axle A, and attach a weight of mass M to the other end of the string (Fig. 3.4b). The length of the string is such that M reaches the floor at the same instant as the string is unwound from the axle.

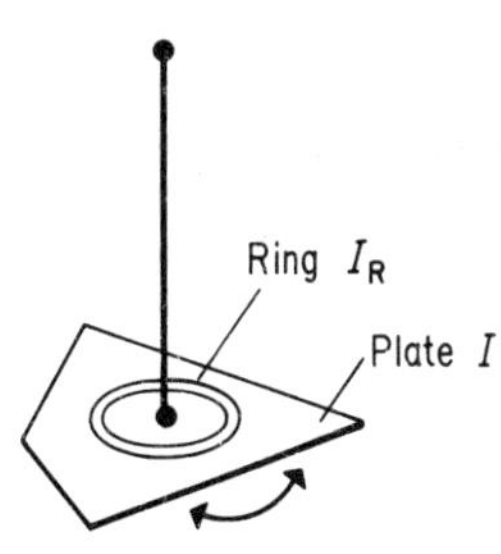

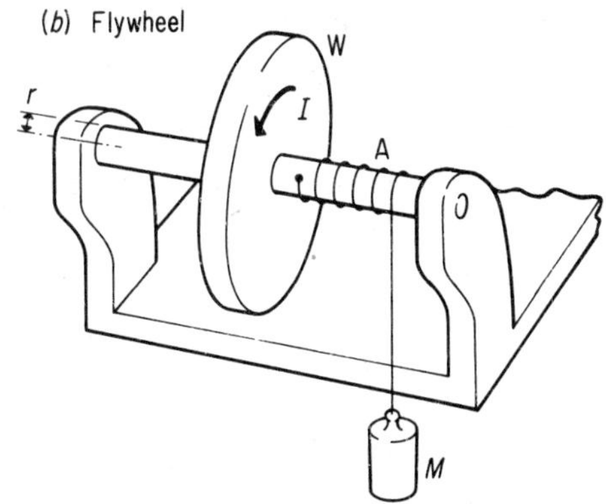

Torsional observations:

$$T_1 = 2\pi\sqrt{\frac{I + I_R}{c}} \quad \cdots (1)$$

$$T_2 = 2\pi\sqrt{\frac{I}{c}} \quad \cdots (2)$$

Allow M to fall to ground:

$$Mgh = \tfrac{1}{2}I\omega^2 + \tfrac{1}{2}Mr^2\omega^2 + nf \cdots (1)$$

$$n_1 f = \tfrac{1}{2}I\omega^2 \quad \cdots (2)$$

FIG. 3.4 Moment of inertia determinations.

Measurements made are: (*i*) the height h dropped, (*ii*) the radius r of the axle, (*iii*) the number of revolutions n of the wheel W while M drops, (*iv*) the further number of revolutions n_1 of the wheel before coming to rest, and t, the time taken, *after M reaches the ground.*

From the energy equation,

$$Mgh = \tfrac{1}{2}Mr^2\omega^2 + \tfrac{1}{2}I\omega^2 + nf,$$

where f is the energy per turn expended against friction. Now $\tfrac{1}{2}I\omega^2 = n_1 f$; and thus

$$Mgh = \tfrac{1}{2}Mr^2\omega^2 + \tfrac{1}{2}I\omega^2\left(1 + \frac{n}{n_1}\right) \quad \cdots \cdots (i)$$

Also, average angular velocity $= \omega/2 = 2\pi n_1/t$, or $\omega = 4\pi n_1/t$. Thus I can be calculated, from (*i*).

EXAMPLE

Explain the meaning of the term *moment of inertia*. Describe in detail how you would find experimentally the moment of inertia of a bicycle wheel about the central line of its hub.

A uniform cylinder 20 cm long, suspended by a steel wire attached to its mid-point so that its long axis is horizontal, is found to oscillate with a period of 2 s when the wire is twisted and released. When a small thin disc, of mass 10 g is attached to each end the period is found to be 2·3 s. Calculate the moment of inertia of the cylinder about the axis of oscillation. (*N.*)

23

The moment of inertia of each disc about the axis of suspension

$$= ma^2 = 0.01 \text{ (kg)} \times 0.1^2 (\text{m}^2) = 10^{-4} \text{ kg m}^2.$$

$$\therefore \quad \text{new moment of inertia} = I + 2 \times 10^{-4},$$

where I is the moment of inertia of the cylinder alone. Since $T = 2\pi\sqrt{I/c}$ generally,

$$\therefore \quad \frac{2.3}{2} = \sqrt{\frac{I + 2 \times 10^{-4}}{I}}$$

$$\therefore \quad I = 2 \times 10^{-4}\left(\frac{2^2}{2.3^2 - 2^2}\right) = 6.2 \times 10^{-4} \text{ kg m}^2.$$

STATICS

Parallelogram of forces. Vector addition. A force is a vector. The sum or resultant of two forces can thus be found by drawing a vector triangle or by drawing a parallelogram of forces. In the latter case the *diagonal* represents the resultant, R.

If P, Q are the two forces, and θ is the angle between them, then

$$R^2 = P^2 + Q^2 + 2PQ \cos \theta.$$

Resolved component. The resolved component of a force X in a direction inclined to it at an angle $\theta = X \cos \theta$; in a perpendicular direction the component is $X \cos (90 - \theta)$, or $X \sin \theta$.

Moment (Torque) of a force about a point or axis O, $=$ force $\times$ perpendicular distance from O to the line of action of the force. Fig. 3.5a *Unit*: newton metre (N m).

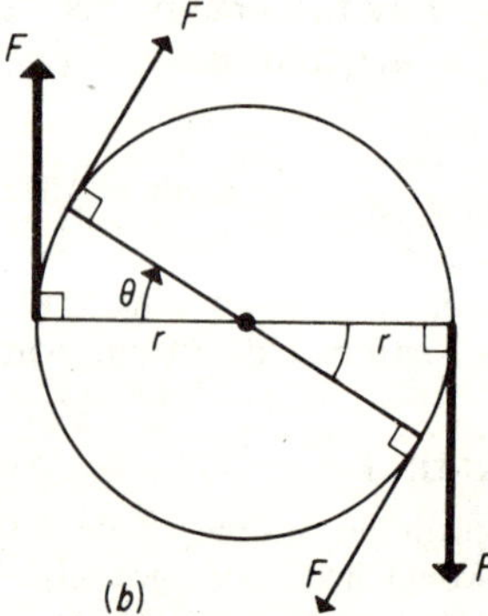

Fig. 3.5 Torque and work.

Moment (Torque) of a couple = one force × perpendicular distance between forces. Fig. 3.5*b*. Unit: newton metre (N m).

Work done by a couple = torque × angle of rotation.

Unit: Work in joule (J) when torque in newton metre and angle in radian.

Three forces in equilibrium (Non-parallel forces). (*i*) The three forces must meet at a point. (*ii*) The forces can be represented by the three sides of a triangle taken in order (*The Triangle of Forces Theorem*).

When three (or more) forces keep a body such as a ladder in equilibrium, the unknown forces can be found, among other methods, as follows:

(*a*) *Resolve all the forces vertically*; the total upward force = the total downward force.

(*b*) *Resolve all the forces horizontally*; the total forces one way = the total forces the other way.

(*c*) *Take moments about a suitable point*, where one or more of the forces can be eliminated. For example, moments taken about the end of a ladder resting on the ground will eliminate the forces at the ground.

Parallel forces in equilibrium. In this case (*i*) the total force in one direction = the total force in the opposite direction, (*ii*) the clockwise moments about any point = the anticlockwise moments about the same point.

The resultant R of a number of parallel forces is the algebraic sum of the forces; and the moment of R about any point = the algebraic sum of the moments of the individual forces about the same point.

Centre of mass. This is the point where the resultant mass of a body acts. Its position, G, may be found by taking moments about a fixed axis, such as AB in Fig. 3.6, for all the distributed masses,

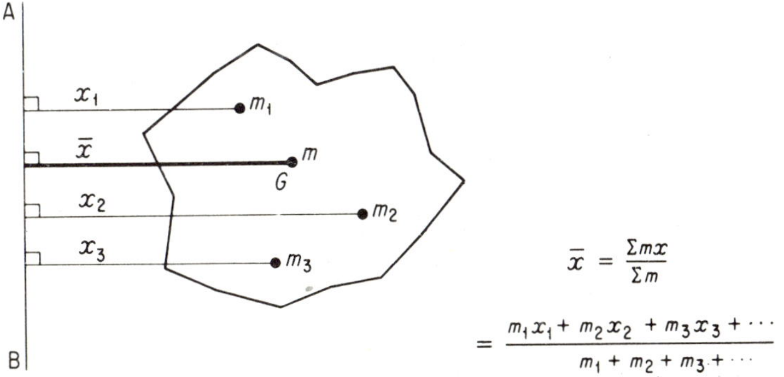

$$\bar{x} = \frac{\Sigma m x}{\Sigma m}$$

$$= \frac{m_1 x_1 + m_2 x_2 + m_3 x_3 + \cdots}{m_1 + m_2 + m_3 + \cdots}$$

FIG. 3.6 Centre of mass.

25

represented by $m_1, m_2, m_3, \ldots$ Then $\bar{x} = \Sigma\, mx/\Sigma m = (m_1 x_1 + m_2 x_2 + m_3 x_3 + \ldots)/(m_1 + m_2 + m_3 + \ldots)$. The c.m. of a *triangular lamina* is two-thirds of the distance along the median from an apex of the triangle (median = the line joining an apex to the mid-point of the opposite side).

The c.m. of the *curved surface* of a hollow cylinder is at the mid-point of its axis and that of a hollow cone is two-thirds of the distance along the axis from the apex of the cone.

The c.m. of a right *solid cone* is three-quarters of the distance along the axis from the apex of the cone.

Centre of gravity. This is the point where the resultant weight of a body acts. It coincides with the centre of mass when the body is small but not when the body is very large.

C.M. of a combined object. The c.m. of an object such as a cylinder partly filled with water can be found by the principle of moments which says:

Moment of resultant mass about any point = algebraic sum of moments of individual masses about this point.

Thus (mass of cylinder and water) $\times \bar{x}$ = mass of cylinder $\times x_1$ + mass of water $\times x_2$, where x is the distance of the c.m. from some point, and x_1, x_2 are the respective distances of the c.m. of the cylinder and water from that point.

Common Balance. (*i*) The fulcrum is an agate wedge on an agate plate, (*ii*) agate wedges support the scale-pans, (*iii*) the c.g. of the beam and pointer is vertically below the fulcrum, to make the arrangement stable, (*iv*) the arms of the lever are equal.

If the arms are unequal, and W_1, W_2 are the respective masses for counter-balance in each scale-pan, the true mass W is given by
$$W = \sqrt{W_1 W_2}.$$

Sensitivity of balance. A balance is sensitive if a small difference

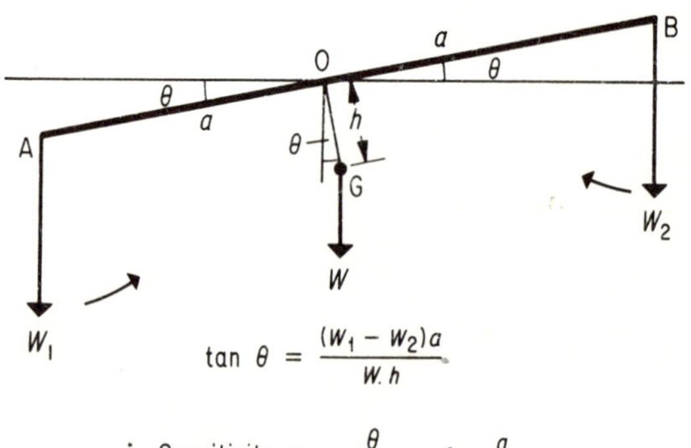

$$\tan \theta = \frac{(W_1 - W_2)a}{W.h}$$

$$\therefore \text{Sensitivity} = \frac{\theta}{W_1 - W_2} \simeq \frac{a}{W.h}$$

Fig. 3.7 Sensitivity of balance.

in weights on the scale-pans causes a large deflection of the beam; sensitivity may be expressed in "divisions per milligram difference".

If θ is the small angle of inclination of the beam when the masses on the scale-pans are W_1, W_2 respectively, h is the depth of the c.g. G of beam and pointer (mass W) below the fulcrum O, then by taking moments about O,

$$\tan\theta = \frac{(W_1 - W_2)a}{Wh},$$

where a is the length of the "arms" of the beam (Fig. 3.7).

Thus high sensitivity if a is large and W, h are small, that is, the arms should be long and the c.g. must be very close to the fulcrum. But a long beam will take a long time to settle down when deflected, as its moment of inertia will be large (p. 22), and a light beam will not be rigid. Thus a compromise is required between sensitivity and design.

Corrections in accurate weighing. The upthrust on a weight or object due to the "buoyancy" of the air depends on the density of air, which is controlled by such factors as the pressure, temperature and relative humidity.

$$\text{True mass} = m_1 \frac{\left(1 - \dfrac{\sigma}{\rho_1}\right)}{1 - \dfrac{\sigma}{\rho}} = m_1\left(1 - \frac{\sigma}{\rho_1} + \frac{\sigma}{\rho}\right), \quad \text{by} \quad \text{binomial}$$

expansion, where m_1 is the mass of the counterpoise weight, σ is the air density, ρ, ρ_1 are the respective densities of the material of the mass weighed and the counterpoise weight.

FLUIDS

Density = mass/volume. SI unit: kg m^{-3}
Density of water at 4°C = 1000 kg m^{-3} (1g cm^{-3}). Density of mercury = 13 600 kg m^{-3} (13·6 g cm^{-3}).

Relative density = mass of substance/mass of equal volume of water

= density of substance/density of water

Relative density (specific gravity) has no units. For mercury, relative density = 13·6.

Pressure in liquids. (1) Pressure, p, is defined as the average *force per unit area* at the place concerned.

$$p = h\rho g,$$

where p is the pressure due to a column h of density ρ and g is the particular acceleration due to gravity. Units: p in newton metre^{-2} (N m^{-2}) when h in metre (m), ρ in kg m^{-3}, $g = 9\text{·}8 \text{ m s}^{-2}$.

(2) Standard atmospheric pressure = pressure due to 760 mm Hg = $0.76 \times 13600 \times 9.8 = 1.013 \times 10^5$ N m^{-2}. 1 atmosphere = 1.01325×10^5 N m^{-2}. 1 bar = 10^5 N m^{-2}, or 10 bar = 1 MN m^{-2} (1 MN = 10^6N) Note that 1 mmHg pressure = 1 torr.

(3) *"Corrected" barometric height.* If this is H_0 at 0°C, then $H_0\rho_0 g = H_t\rho_t g'$, where g is the acceleration due to gravity at sea-level, latitude 45°, and g' is that where the barometer height H_t is read. Thus

$$H_0 = H_t \frac{\rho_t}{\rho_0} \times \frac{g'}{g}.$$

Since $\rho_0/\rho_t = 1+\gamma t$, where γ is the cubic expansivity of mercury, and $H_t = H'(1+\alpha t)$, where H' is the reading in the metal scale assumed accurate at 0°C and α is the linear expansivity of the metal,

$$H_0 = \text{"corrected" height} = H'\frac{1+\alpha t}{1+\gamma t} \cdot \frac{g'}{g} = H'[1-(\gamma-\alpha)t]g'/g.$$

The magnitude of g'/g is obtained from standard tables. Correction is also required for the surface tension of mercury for further accuracy.

Bernouilli's Principle. For an incompressible, non-viscous fluid moving with a uniform or streamline motion, then, at any point on a streamline, *the sum of the pressure, kinetic energy per unit volume and potential energy per unit volume is always a constant.*

Proof. Work done by excess pressure per unit volume of fluid (p_1-p_2) = gain in K.E. per unit volume $(\frac{1}{2}\rho v_2{}^2 - \frac{1}{2}\rho v_1{}^2)$ + gain in P.E. per unit volume $(\rho gh_2 - \rho gh_1)$

$$\therefore \quad p_1+\tfrac{1}{2}\rho v_1{}^2+\rho gh_1 = p_2+\tfrac{1}{2}\rho v_2{}^2+\rho gh_2$$

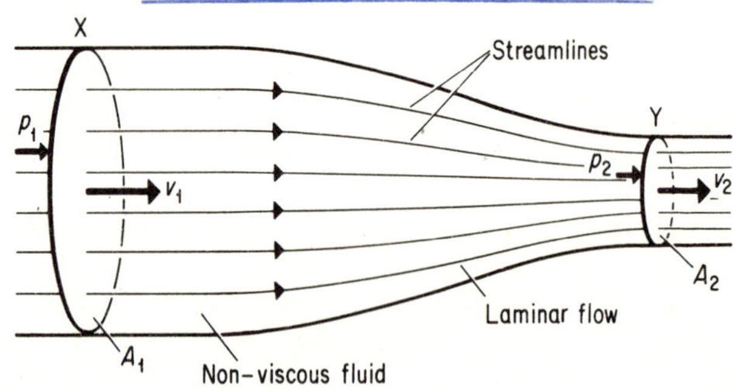

$$p_1 > p_2, \quad v_1 < v_2,$$

Volume per second = $A_1 v_1 = A_2 v_2$,

$$p_1 - p_2 = \tfrac{1}{2}\rho(v_2^2 - v_1^2)$$

FIG. 3.8 Bernoulli's principle.

Bernoulli's Principle is thus a consequence of the principle of the conservation of energy.

Along a horizontal pipe, suppose the area A_1 of section X $= 10\ \text{cm}^2$ and area A_2 of section Y $= 5\ \text{cm}^2$. Then velocity of liquid at Y, $v_2 = 2 \times$ velocity of liquid at X, v_1. Pressure difference between X and Y $= p_1 - p_2 =$ gain in K.E. per unit volume $= \frac{1}{2}\rho\, v_2{}^2 - v_1{}^2)$. Fig. 3.8.

For horizontal flow with the above conditions, the Principle shows that the pressure is high in the fluid where its velocity is low, and vice-versa. Fast-moving air blown between two near parallel cards result in lower pressure here than on the other side, so that the two cards are forced together by the excess pressure.

Applications. 1. *Aerofoil lift* is due to faster airflow over the top surface than over the bottom surface. The greater pressure below produces a net uplift.

2. Fast-moving air is blown across the top of a *scent-spray* bottle when the bulb is squeezed. This reduces the pressure here and the atmospheric pressure on the liquid in the bottle forces it out.

3. *Torricelli's theorem.* The velocity of liquid emerging at a depth h from a hole in a wide vessel open to the air is equal to $\sqrt{2gh}$ approximately. This is the velocity it would have had if falling a height h freely under gravity. Proof: Apply Bernoulli's principle to approximate streamline flow from (i) top of liquid to (ii) hole. Then, approximately, if B is atmospheric pressure,

$$\text{(i)}\ B + h\rho g = \text{(ii)}\ B + \tfrac{1}{2}\rho v^2, \quad \text{or} \quad v^2 = 2gh.$$

EXAMPLE

A non-viscous liquid of density $1200\ \text{kg m}^{-3}$ moves with steady streamline flow at $1\cdot 2 \times 10^{-2}\ \text{kg s}^{-1}$ through a horizontal tube T of uniform cross-sectional area $20\ \text{cm}^2$. Calculate the liquid speed. Further along, the tube narrows slowly to a cross-sectional area X of $10\ \text{cm}^2$. Calculate the new speed through X and the difference in pressure between X and T, assuming streamline flow throughout.

(*i*) Mass of liquid per second crossing section of T
$$= \text{cross-section area} \times \text{speed} \times \text{density}$$
$$\therefore \quad 1\cdot2 \times 10^{-2} = 20 \times 10^{-4} \times v \times 1200, \quad \text{or} \quad v = \tfrac{1}{200}\ \text{m s}^{-1}.$$

(*ii*) $\qquad$ Speed through X $= 2 \times$ speed through T $= \cdot\tfrac{1}{100}\ \text{m s}^{-1}$

(*iii*) Difference in pressure $=$ gain in P.E. per unit volume $= \tfrac{1}{2}\rho(v_2{}^2 - v_1{}^2)$

$$= \tfrac{1}{2} \times 1200 \times \left(\frac{1}{100}^2 - \frac{1}{200}^2\right) = 4\cdot5 \times 10^{-2}\ \text{N m}^{-2}$$

Electrostatics

---

4. CHARGES. GENERATORS

Negative and positive charges. By definition, an ebonite rod rubbed with fur has a negative charge; a glass rod rubbed with silk has a positive charge. A polythene rod, rubbed by a duster, has a negative charge; a cellulose acetate strip has a positive charge.

Like charges repel; unlike charges attract.

Conservation of charge. Friction does not create electricity or charges. Charges are present in the atoms of the rod X and the rubbing material Y before rubbing takes place. The friction results only in the transfer of electrons from X to Y, or from Y to X, according to the nature of the materials. Since the total negative and positive charges in neutral atoms are equal, there is now a surplus negative charge on X say, leaving an *equal* surplus positive charge on Y. The total charge on X and Y is zero, the same as the case before rubbing. Similarly, if charges are transferred between atoms, or between charged metal plates, the total charge before transfer $=$ the total charge after transfer.

In insulators, the electrons in atoms are "bound" to the nucleus. In metals, there are one or more "free" electrons per atom. All charges are made up of units of e, the numerical value of the charge on an electron, that is, charges are "granular".

Gold-leaf electroscope. Consists of (*i*) a gold leaf, or leaves, at the end of a metal stem S with a disc or cap, (*ii*) a metal box surrounding the leaves and insulated from the stem, (*iii*) plane glass windows for observing the divergence of the leaf. The metal box is usually earthed, and thus shields the leaf from outside electrical influences other than those undergoing test.

Experiments with gold-leaf electroscope. (1) *Testing sign of charges.* Give the leaf a negative charge by touching it with a suitable charged rod or charging it by induction. If an unknown charge A causes increased divergence, then A has a negative charge. If the leaf has a diminished divergence, A has positive charge or may be uncharged. To test, discharge the leaf, give it a positive charge by contact, and bring up A. If the leaf diverges further, A has a positive charge. Increased divergence of the leaf is the only sure test of charge.

(2) *Testing magnitudes of charges.* Place a can on top of the electro-

scope cap. Insert a charge into the can and touch the bottom with the charge. *All* the charge is then given to the electroscope, see ice-pail experiment, p. 32, so that the divergence of the leaf is a measure of its magnitude.

(3) *Testing conducting powers of different threads.* Insulate the case, and connect the cap to the case by a thread. Charge the cap, and observe the rate of fall of the leaf. The thread corresponding to the greatest rate of fall has the greatest conducting power.

(4) *Testing equal charges when electrifying by friction.* Holding fur by an insulator, rub an ebonite rod with the fur, and bring fur and rod together near the electroscope cap. No divergence implies equal charges.

(5) *Demonstrating charges from battery supply and polarity sign.* (Care.) Connect a high d.c. supply, with a protective or built-in resistance of some megohms between the cap and case of an electroscope. Find the sign of the charge on the leaf with a positive or negative charged rod. This gives the sign of the two poles of the supply.

The divergence of the leaf is a measure of the potential difference between the cap and the case of the electroscope, not of the charge on the cap. This can be demonstrated by connecting the cap to the case, insulating the latter, and then touching the cap with a charged rod so that the cap acquires a charge. No divergence occurs. When the case is insulated, and the cap is earthed, the leaf diverges when a charged rod is brought near to the case.

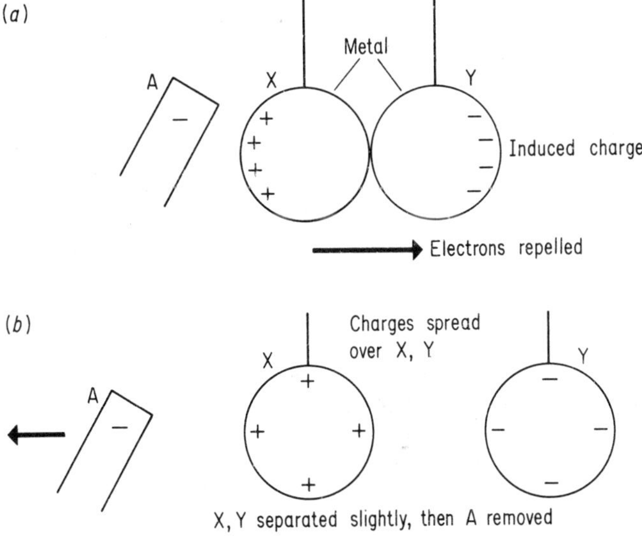

Fig. 4.1 Induced charges.

Induced charges. A charge may be *induced*; the *inducing charge* is then left unaffected. In Fig. 4.1*a*, the charge A repels electrons from the metal sphere X to sphere Y, leaving X positively charged. On separating X, Y and then removing A, induced charges appear on X and Y. Fig. 4.1*b*.

Charging electroscope by induction. (Fig. 4.2). Bring the charged rod R near to the cap (whereupon the leaf diverges); touch the cap (the leaf closes); remove the finger (the leaf remains closed); finally

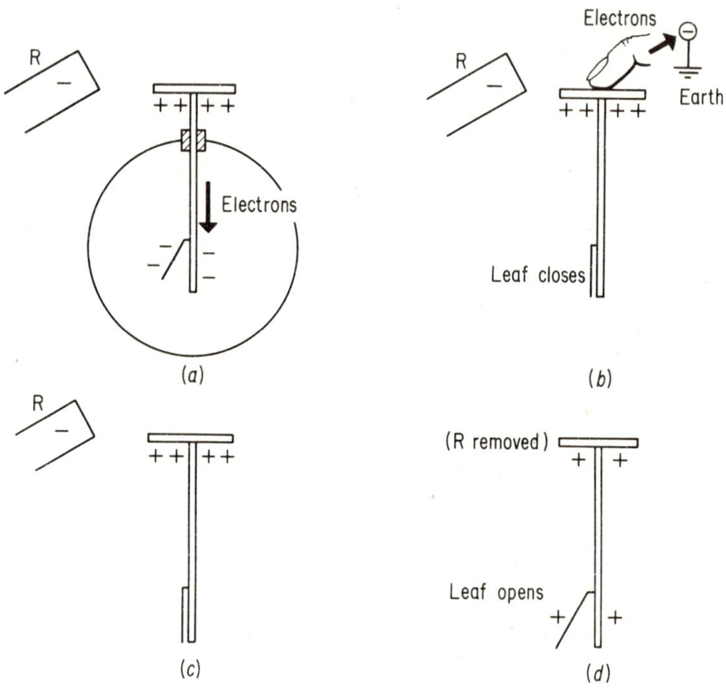

FIG. 4.2 Charging electroscope by induction.

remove the rod (the leaf opens). The induced charge on the leaf is opposite to the inducing charge on the rod.

Faraday's ice-pail experiment on induced charges.   An insulated ice-pail B was connected to a gold-leaf electroscope (Fig. 4.3). (*i*) On placing a charge A inside the pail B the leaf diverged; the divergence was constant when A was moved about near the bottom of B, taking care not to touch B. When A was withdrawn from B the divergence of the leaf diminished to zero.

(*ii*) On placing A again in B the leaf diverged, and the divergence

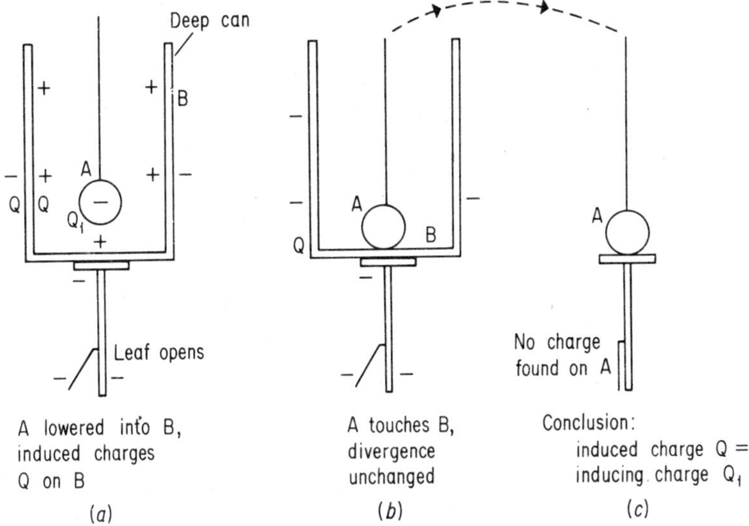

A lowered into B,
induced charges
Q on B

(a)

A touches B,
divergence
unchanged

(b)

Conclusion:
induced charge Q =
inducing charge Q₁

(c)

FIG. 4.3 Faraday's ice-pail experiment.

was unaltered when A touched the inside of B. But on withdrawing A and testing, it was found to have no charge.

Conclusions. Since the leaf closed again on withdrawing A from B in (*i*), it follows that the induced negative and positive charges on the electroscope and pail are *equal.* Otherwise there would be a surplus charge when A was withdrawn, and the leaf would then diverge.

Since A has no charge left when it touches the interior of B, from (*ii*), we conclude that *the inducing charge on* A *is equal and opposite to the induced charge on* B. Moreover, as A has not collected any of the remaining charge on B, it follows that *the charge on* B *resides on the exterior.*

Note. If A has a positive charge and is placed inside B, an induced negative appears on the inside of B, and an equal positive charge appears on the outside of B and on the cap and leaf of the electroscope.

Surface density. The surface density of charge is the *charge per unit area* on the part of the conductor concerned. The surface density is tested by touching the conductor with a *proof-plane* and transferring the charge collected to the inside of a can on top of a gold-leaf electroscope. The divergence of the leaf is a measure of the surface density. The surface density is greatest at those parts of the conductor which have the greatest curvature, such as points. See Fig. 4.4.

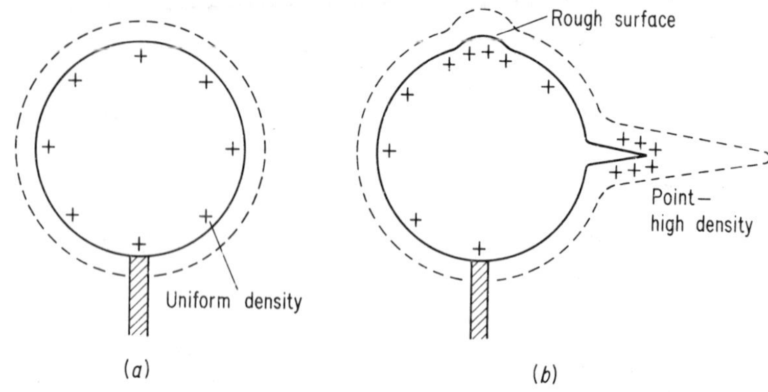

FIG. 4.4 Variation of surface-density of charge.

Location of charge on a conductor. By testing the interior of a large hollow charged conductor with a proof-plane, it can be shown that no charge is present *inside* a conductor. All the charge resides on the outer surface. Faraday demonstrated this (*i*) by turning a charged conducting butterfly-net inside out and testing for charge with a proof-plane each time, (*ii*) by building a large metal box and then sitting inside with an electroscope. Violent electric discharges outside the box caused him no discomfort, and the electroscope was unaffected. If there were charges in the box, the electroscope would be affected.

Action at points. A bent metal rod, pointed at both ends, can be balanced in the middle on the metal dome of a Van de Graaff generator. When the machine is started, the rod rotates in the opposite direction to the pointed ends. The rotation is due to the "action at points". The point of the charged rod has a high density of charge.

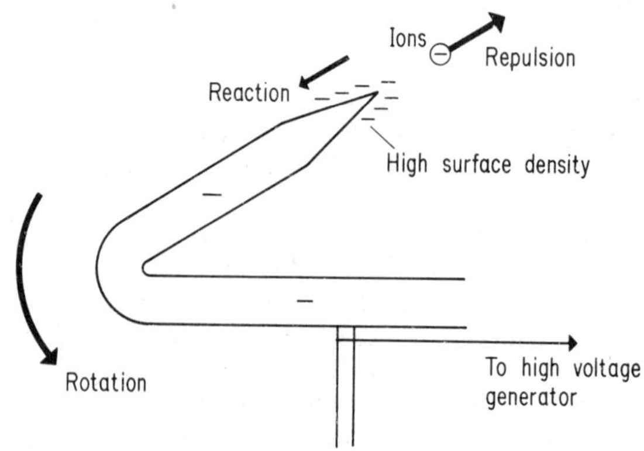

FIG. 4.5 Action at points.

The air in front of it is then ionised. The negative charges on the ions
are repelled, so that they stream away from the point ("electric
wind"). See Fig. 4.5. By the law of action and reaction, the rod is
repelled in the opposite direction and hence rotates. The positive
charges on the other ions are attracted towards, and partly discharge, the point. In effect, then, a pointed conductor loses charge
to the air by the "action of points".

Electrophorus. This is a simple electrostatic generator. Action:
Place the metal disc B on A, a charged insulator. Fig. 4.6a. Induced

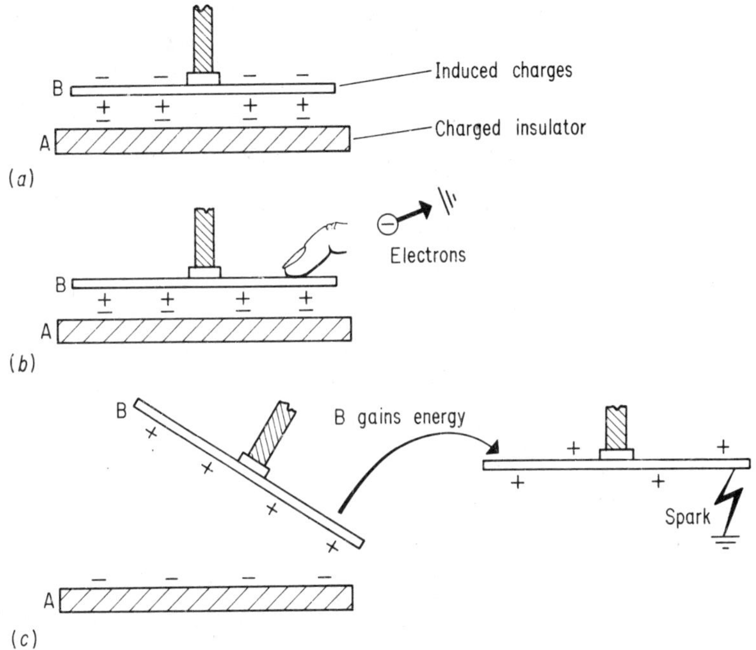

FIG. 4.6 Electrophorus.

charges appear on B. Touch B, electrons are repelled to the earth.
Fig. 4.6b. Remove B from A. Fig. 4.6c. It has an induced positive
charge. Note that B gains energy on removal from A, as work is
done against the force of attraction. This is shown by a spark when
B is discharged.

Van de Graaff generator. This produces a very high voltage by
electrostatic means. It consists basically of (i) a continuous silk belt S
running round a high and low pulley, H, L respectively, (ii) a
polished metal hemisphere M, the high-voltage terminal, at the top
of a pyrex column surrounding H, (iii) a battery of a few thousand
volts at the bottom, with one terminal connected to a set of points P

and the other terminal earthed, (iv) a motor driving the pulley L so that the band rotates, (v) a set of points A, opposite the band at the top, which is connected to M. (Fig. 4.7).

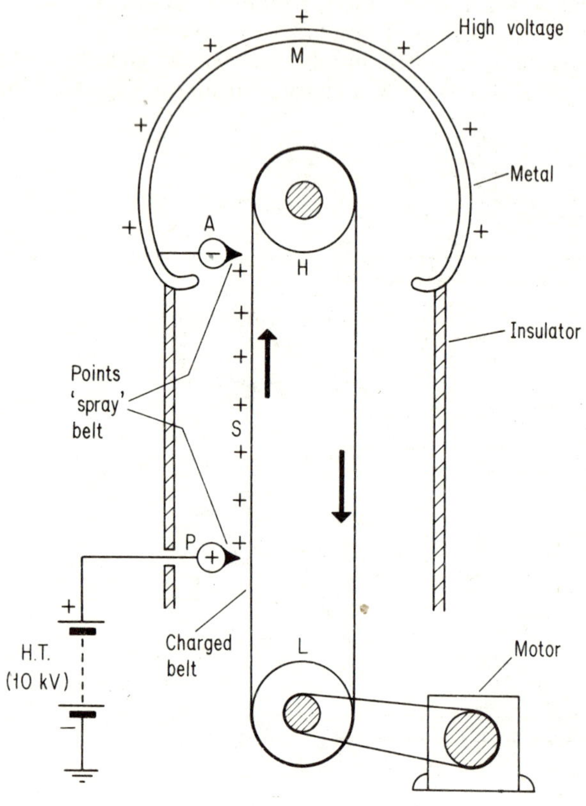

FIG. 4.7 Van de Graaff generator.

Action. P "sprays" the belt at the bottom with say positive charge, by the action at points. The belt rises with the charge, and reaches the points A. Here a negative charge is induced on A and a positive charge on M. A now "sprays" the belt immediately in front of it with negative charge, by the action at points, and neutralizes the positive charge on the belt. The action continues with M gaining an ever-increasing charge; the potential of M grows accordingly, being limited only by the size of M and the breakdown of the air. Several million volts can be reached.

5. ELECTRIC FORCE. POTENTIAL

Force between two charges.

$$F = \frac{Q_1 Q_2}{4\pi\varepsilon_0 r^2} \quad \text{(vacuum)} \quad \text{or} \quad \frac{Q_1 Q_2}{4\pi\varepsilon r^2} \quad \text{(dielectric) Fig. 5.1}a.$$

(i) F in newtons (N) when Q_1, Q_2 in coulombs (C), r in metres and $\varepsilon_0 = 8\cdot85 \times 10^{-12}$ farad metre^{-1} (approx.) = permittivity of a vacuum. (ii) $\varepsilon = \varepsilon_r\varepsilon_0$, where ε_r is the relative permittivity (dielectric constant) of the medium and is a number or ratio, so that it has no units.

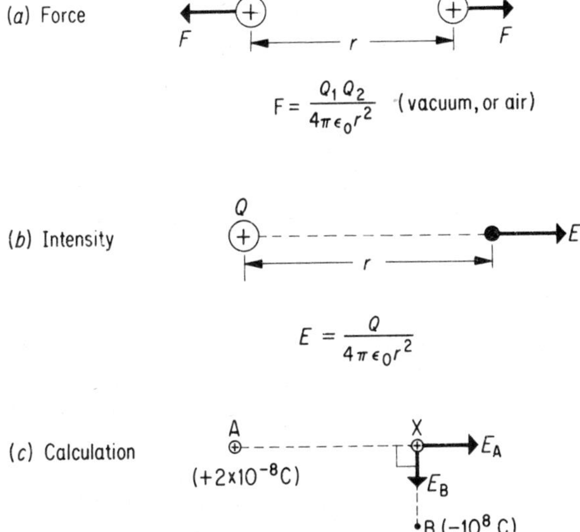

FIG. 5.1 Force between charges. Electric intensity.

(iii) To simplify calculations, note that $1/4\pi\varepsilon_0 = 9 \times 10^9$ to a good approximation. Thus the force of repulsion between two small positive charges of 10^{-9} and 2×10^{-9} C respectively placed 5 cm apart

$$= \frac{1}{4\pi\varepsilon_0}\frac{Q_1 Q_2}{r^2} = 9 \times 10^9 \times \frac{10^{-9} \times 2 \times 10^{-9}}{(5 \times 10^{-2})^2} = 7\cdot2 \times 10^{-6}\text{N}$$

Electric Intensity. The electric intensity E at a point in a field is defined as the *force per coulomb on a very small test charge placed at the point*. The direction of E is taken as the direction of movement of a positive test charge.

Unit of E: newton coulomb (N C^{-1}) or volt metre^{-1} (V m^{-1})—see p. 42.

The force on a charge Q at a point where the intensity is E is given by $F = EQ$.

Units: F in newtons when E in newton coulomb^{-1} and Q in coulomb.

At a distance r from a small (point) charge Q, $E = Q/4\pi\varepsilon_0 r^2$. Fig. 5.1$b$.

EXAMPLE

Find the resultant intensity at a point X distant 10 cm from a point charge A of $+2 \times 10^{-8}$ C and 5 cm from a point charge B of -10^8 C, where angle AXB $= 90°$. Fig. 5.1c.

Imagine a positive test charge at X. Then, using $1/4\pi\varepsilon_0 = 9 \times 10^9$,

$$E_A = \frac{Q}{4\pi\varepsilon_0 r^2} = \frac{9 \times 10^9 \times 2 \times 10^{-8}}{(10 \times 10^{-2})^2}$$

$$= 18 \times 10^3 \text{ N C}^{-1}, \quad \text{along AX}$$

$$E_A = \frac{9 \times 10^9 \times 10^{-8}}{(5 \times 10^{-2})^2} = 36 \times 10^3 \text{ N C}^{-1}, \quad \text{along XB}$$

$$\therefore \quad \text{resultant } E = \sqrt{(E_A^2 + E_A^2)} = 10^{-3}\sqrt{(18^2 + 36^2)}$$

$$= 4 \times 10^{-2} \text{ N C}^{-1} \quad \text{(approx.)}$$

Electric flux and flux density. The pattern of electric flux in the neighbourhood of some charged conductors are shown in Fig. 5.2a, b, c and Fig. 5.4. Note (i) the radial flux round a point charge and a charged sphere and the contrasting parallel flux between two parallel plates with opposite charges in Fig. 5.4, (ii) the absence of flux and hence field inside a charged conductor. In drawing fields, note that the lines leave the surfaces of conductors normally.

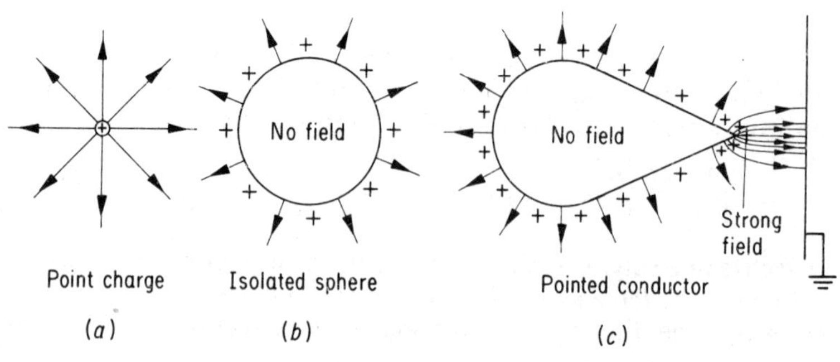

FIG. 5.2 Electric flux round charged conductors.

The electric intensity E at a point in a field may be defined as the *flux density* there. The flux density is equal at points at the same distance from a point charge. The flux density is equal and in the same direction for most of the region between the two parallel plates in Fig. 5.4. The flux density is high round the point of a pear-shaped conductor, Fig. 5.2c, so that the intensity E is high at this place.

Some intensity values. (1) *Charged hollow or solid spherical conductor.* Outside: $E = Q/4\pi\varepsilon_0 r^2$, where r is the distance from the centre. At the surface: $E = Q/4\pi\varepsilon_0 a^2$, where a is the radius of the sphere. Fig. 5.3a. Inside: $E = 0$—this result applies to any shape of conductor (see Faraday's demonstration, p. 33, Fig. 4.3).

(2) *Charged plane conductor.* Just outside the conductor, where the flux is normal to the plane, $E = \sigma/\varepsilon_0$ in a vacuum or air. σ is the "charge density", measured in coulomb metre^{-2}, on the conductor.

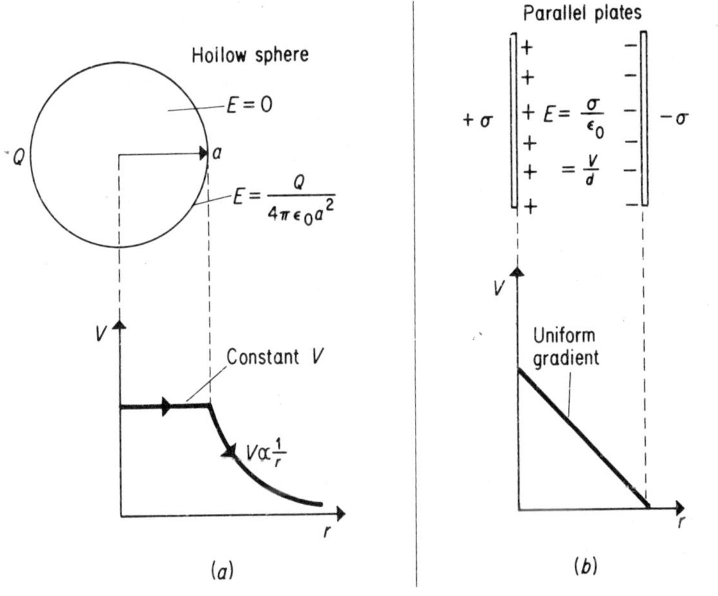

FIG. 5.3 Electric intensity and potential.

(3) *Between two long parallel metal plates with equal charges,* e.g. parallel plate capacitor, $E = \sigma/\varepsilon$, where ε is the permittivity of the dielectric completely filling the space between the plates. Fig. 5.3b. Owing to the non-uniformity of the field at the edges ("edge effect"), E is less at the edges, Fig. 5.4, and "guard-rings" are therefore used in experiments with parallel plates.

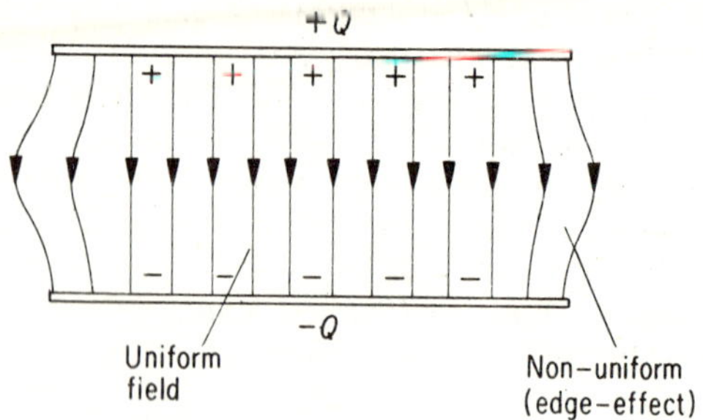

FIG. 5.4 Field between parallel charged plates.

Note that the formula for E due to a point charge or outside a charged sphere has "4π" in the denominator; this is due to the radial symmetry of the field. In contrast, the formula for E between two charged parallel plates has no "π" in the denominator; this is due to the uniform field (straight parallel lines).

Potential. The *electric potential* at a point is equal to the work done per coulomb in bringing a small positive charge from infinity to that point. Earth potential is the practical "zero" of potential.

The *potential difference (p.d.)* between two points is equal to the work done per coulomb in taking a small positive charge between the two points.

The *volt* is the p.d. when 1 joule of work is done in taking 1 coulomb of charge between the points.

If V_1 is the potential at a point X, V_2 at a point Y, then the work done W in taking a charge Q between the points is given by

$$W = Q (V_2 - V_1)$$

Potential values. (1) The potential V at a point distant r from a *point charge Q* in a vacuum or air $= Q/4\pi\varepsilon_0 r$.

Note that r must be in metres and Q in coulombs, when V is in volts. Approximately, $1/4\pi\varepsilon_0 = 9 \times 10^9$.

Proof:
$$V = \int_\infty^r -\frac{Q}{4\pi\varepsilon_0 x^2} \cdot dx = \frac{Q}{4\pi\varepsilon_0}\left[\frac{1}{x}\right]_\infty^r = \frac{Q}{4\pi\varepsilon_0 r}$$

If Q is negative in sign, then V is negative.

(2) *Outside* a hollow or solid charged spherical conductor in air, $V = Q/4\pi\varepsilon_0 r$, where r is the distance of the point concerned from the centre.

At the *surface* of a charged sphere, $V = Q/4\pi\varepsilon_0 a$, where a is the radius of the sphere.

Inside a charged sphere, $V = Q/4\pi\varepsilon_0 a$, where a is the radius. This is the potential at *all* points inside a hollow charged sphere, so that the potential is the same as that at the surface. The interior of a hollow charged sphere is thus an *equipotential volume*. The surface of a charged conductor is an *equipotential surface*. See Fig. 5.3*a*.

Note that the potential V at a point is independent of the path taken to reach that point. Thus V is a *scalar* quantity. The total potential at a point due to several charges is hence the algebraic sum of the potentials due to each charge.

EXAMPLES

The charges at the four corners of a rectangle are respectively $+10, -8, -4$ and $+16 \times 10^{-9}$ C. What is the potential at a point P whose distances from the four charges are respectively 5, 4, 4, 8 cm.

$$V_P = \frac{10^{-9}}{4\pi\varepsilon_0}\left(\frac{+10}{5\times10^{-2}} - \frac{8}{4\times10^{-2}} - \frac{4}{4\times10^{-2}} + \frac{16}{8\times10^{-2}}\right)$$

$$= 9\times10^9\times10^{-9}\,(+200-200-100+200)$$

$$= +900 \text{ V}$$

Relation between electric intensity and potential difference.   The electric intensity, E, is the force per coulomb. Thus the p.d., V, between two points a small distance x apart $=$ work done per coulomb in taking a charge between them $= E.x$, where E is the average intensity between the points.

$$\therefore \quad E = \frac{V}{x} = \text{potential gradient between points}$$

In terms of the calculus,

$$E = -\frac{dV}{dx} = \textit{potential gradient}$$

numerically at the point concerned; the minus sign signifies that the potential decreases in the direction of the field. Thus due to a small positive charge Q, $V = Q/4\pi\varepsilon_0 r$ (p. 40). Hence, at the same point, $E = -dV/dr = Q/4\pi\varepsilon_0 r^2$. Note that the potential gradient is constant between charged close parallel plates, except at the edges See Fig. 5.4.

Since $E = 0$ inside a spherical hollow charged conductor (p. 39), it follows that the potential gradient $= 0$. Hence the potential is constant at all points inside the conductor and equal to the potential at the surface. For the same reason, metal cages on insulating columns provide electrostatic shielding inside them. Fig. 5.5.

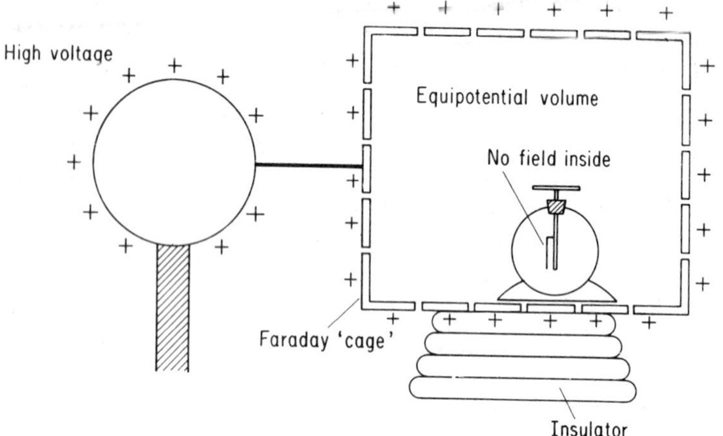

FIG. 5.5 Electrostatic shielding.

Units: Since V is in volts and x in metres, then E can be expressed in *volt metre*$^{-1}$, V m^{-1}. *The unit "volt metre*$^{-1}$*" is the same as the unit "newton coulomb*$^{-1}$*" used previously for E.* This follows from:

$$1 \text{ newton} \times 1 \text{ metre} = 1 \text{ joule} = 1 \text{ coulomb} \times 1 \text{ volt}$$

EXAMPLES

(1) Find the radius of an oil drop held in equilibrium between two parallel plates 5 mm apart when the p.d. is 3kV, if the drop carries a charge $-2e$. (Assume $e = 1 \cdot 6 \times 10^{-19}$ C, density of oil $= 800$ kg m^{-3}, $g = 10$ m s^{-2}).

Force on drop due to gravity $=$ upward force due to electric field (EQ)

$$\therefore \quad mg = E \times 2e$$

$$\therefore \quad \tfrac{4}{3}\pi a^3 \rho g = E \times 2e.$$

$$\therefore \quad a = \left[\frac{3E \times 2e}{4\pi \rho g}\right]^{1/3}$$

Now $E =$ potential gradient $= 3000$ V$/5 \times 10^{-3}$ m $= 6 \times 10^5$ V m^{-1}.

$$\therefore a = \left[\frac{3 \times 6 \times 10^5 \times 2 \times 1 \cdot 6 \times 10^{-19}}{4\pi \times 800 \times 10}\right]^{1/3}$$

$$= 1 \cdot 8 \times 10^{-6} \text{ m.}$$

(2) The breakdown potential gradient in air is 3×10^6 Vm^{-1}. Calculate the potential to which a sphere of radius 2 metres can be raised before it begins to lose its charge to the air.

If Q is the maximum charge on the sphere, potential gradient at surface $=$ intensity $E = Q/4\pi\varepsilon_0 r^2$.

$$\therefore \quad 3 \times 10^6 = \frac{Q}{4\pi\varepsilon_0 \times 2^2}$$

$$\therefore \quad Q = 4\pi\varepsilon_0 \times 12 \times 10^6 \ \text{C} \ . \ . \ . \ . \ . \ (i)$$

The potential V of the sphere $= Q/4\pi\varepsilon_0 r$.   From (i),

$$\therefore \quad V = \frac{12 \times 10^6}{2} = 6 \times 10^6 \ \text{V}.$$

Potential variations. Equipotentials. Fig. 5.6a, b, c shows the potential variation from the centre A of a charged sphere before and after a long conductor B is placed near it and earthed. Note that the potential is constant along a conductor.

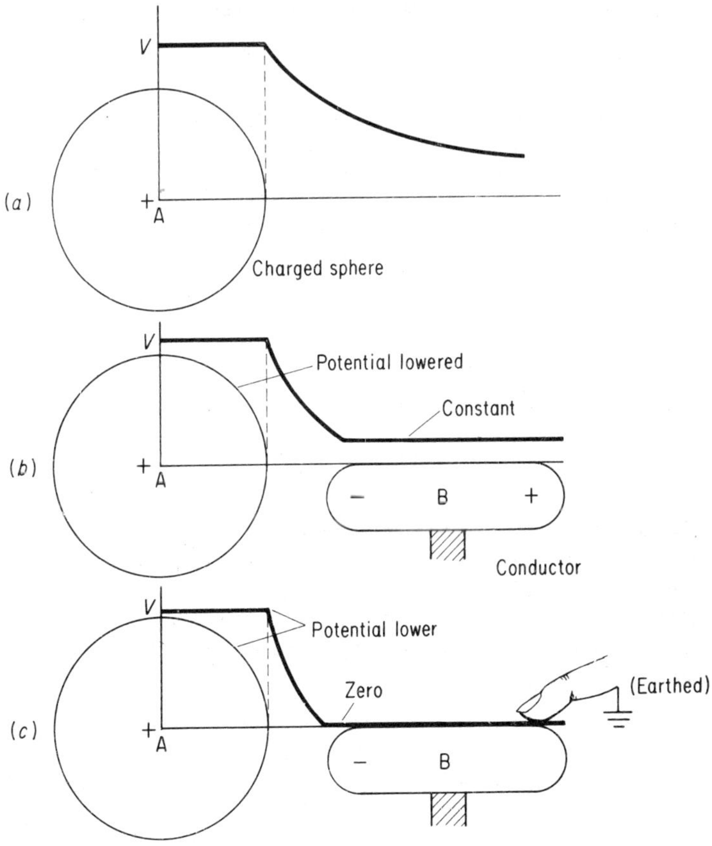

FIG. 5.6 Potential variations.

"Equipotentials" are normal to the flux lines since no work is done in moving along them. Fig. 5.7a. In cathode-ray tubes, electrons are focused by an "electron lens"—the electrons tend to move normally to the equipotentials in the field. Fig. 5.7b.

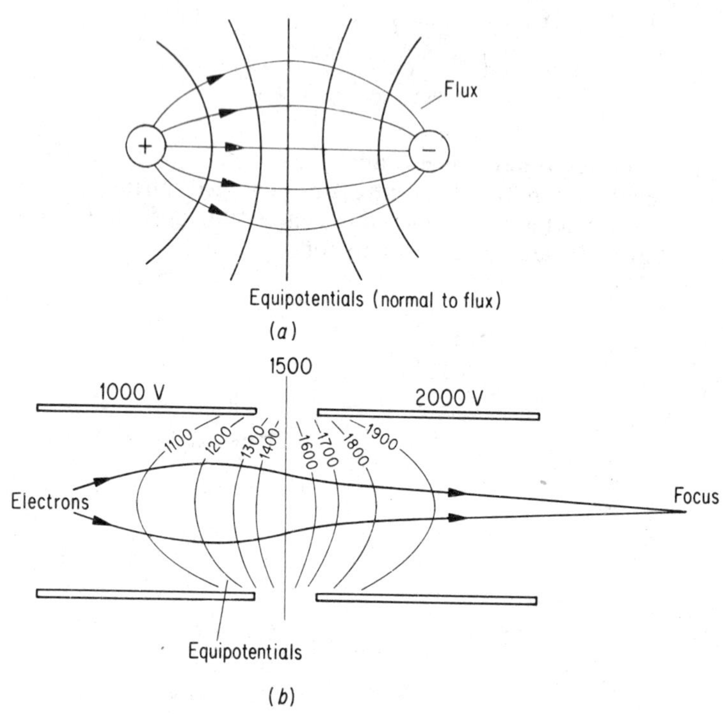

Fig. 5.7 Equipotentials. Electron lens.

Gauss's theorem. This states: The total normal electric flux of E over a closed surface $= Q/\varepsilon$, where Q is the charge in coulombs enclosed by the surface and ε is the permittivity of the medium.

Intensity due to charged shere. (1) *Outside sphere.* Draw a concentric sphere of radius r through the point and apply Gauss's theorem. Then,

$$\text{total normal flux} = \text{area} \times \text{flux-density} = 4\pi r^2 \times E$$

$$\therefore \quad E \times 4\pi r^2 = Q/\varepsilon$$

$$\therefore \quad E = \frac{Q}{4\pi\varepsilon r^2}$$

(2) *Inside sphere.* Draw a concentric sphere through the point concerned. No charge is enclosed. Hence $4\pi r^2 \times E = 0$, or $E = 0$.

Intensity due to infinitely-long charged plane conductor. Draw a pill-box cylinder to enclose a circular area A of the conductor. Both sides of the conductor carry charge. Hence the charge enclosed is $2\sigma A$, where σ is the charge surface-density. If the cylinder is drawn close to the plane and *all* the flux passes through the plane circular ends of area A, then

$$\text{total normal flux} = 2A . E = Q/\varepsilon = 2\sigma A/\varepsilon.$$

$$\therefore \quad E = \sigma/\varepsilon.$$

Units: E in V m^{-1} (or N C^{-1}) when σ in coulomb metre^{-2} and $\varepsilon = \varepsilon_r\varepsilon_0 = \varepsilon_r \times 8{\cdot}85 \times 10^{-12}$.

Force per unit area (stress) on surface. The charge on each part of a plane conductor in air is repelled by the charge on the remainder. The intensity on a part X of the conductor due to the remainder $= \sigma/2\varepsilon_0$, or half the total intensity. The charge per unit area at X $= \sigma$.

$$\therefore \quad \text{force per unit area} = \sigma \times \sigma/2\varepsilon_0 = \sigma^2/2\varepsilon_0$$

Units: The force per unit area is in N m^{-2} when σ is in C m^{-2} and $\varepsilon_0 = 8{\cdot}85 \times 10^{-12}$.

6. CAPACITORS

Capacitance. The capacitance, C, of a *capacitor* is the ratio Q/V, where Q is the charge on either plate when a p.d. V is applied.

The capacitance, C, of an *isolated conductor* is the ratio Q/V, where Q is the charge on the conductor and V its potential relative to earth. (The "earth" is thus the second "plate" of the capacitor.)

Units: C is in farads (F), when Q is in coulombs (C) and V in volts (V).

1 microfarad (μF) $= 10^{-6}$ F; it is the capacitance of a capacitor if the charge on it is 1 microcoulomb (μC) when a p.d. of 1 volt is applied.

1 picofarad (pF) $= 10^{-12}$F.

Dielectric constant (or **relative permittivity**) of a medium, ε_r, $= C/C_0$, where C is the capacitance of a capacitor with the medium completely filling the space between its plates and C_0 is the capacitance of the same capacitor with a vacuum (or air) between its plates.

Values of C. For an *isolated sphere* in air, $C = 4\pi\varepsilon_0 r$, where r is the radius in metres. Fig. 6.1a.

Proof: $V = Q/4\pi\varepsilon_0 r$ at the surface (p. 40). Hence $Q/V = 4\pi\varepsilon_0 r = C$.

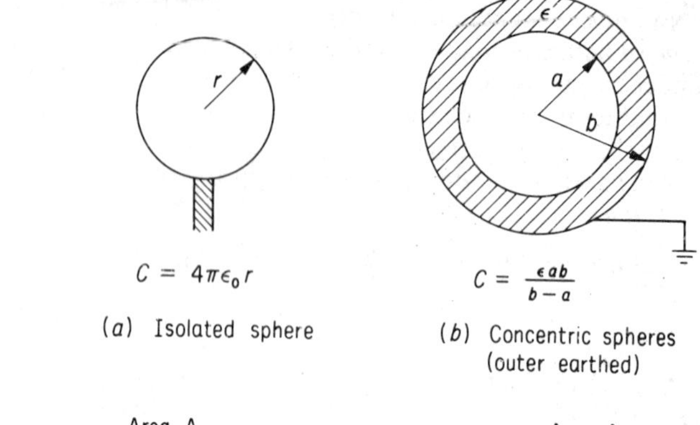

$$C = 4\pi\epsilon_0 r$$

(a) Isolated sphere

$$C = \frac{\epsilon ab}{b-a}$$

(b) Concentric spheres (outer earthed)

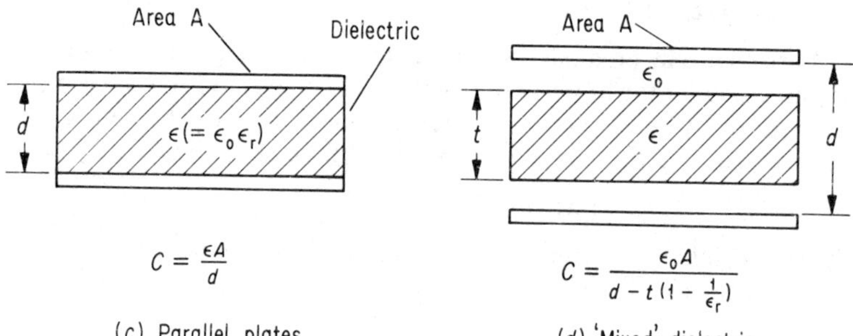

$$C = \frac{\epsilon A}{d}$$

(c) Parallel plates

$$C = \frac{\epsilon_0 A}{d - t\left(1 - \frac{1}{\epsilon_r}\right)}$$

(d) 'Mixed' dielectric

FIG. 6.1 Capacitance values

EXAMPLE

Find the capacitance of an isolated metal sphere of radius 5 cm. Using $1/4\pi\varepsilon_0 = 9 \times 10^9$ (approx.),

$$\therefore \quad C = 4\pi\varepsilon_0 r = \frac{1}{9 \times 10^9} \times 5 \times 10^{-2} = 5{\cdot}5 \times 10^{-12}\ \text{F} = 5{\cdot}5\ \text{pF}.$$

For a *concentric spherical capacitor* in air, $C = 4\pi\varepsilon_0 ab/(b-a)$, where a, b are the radii in metres. Fig. 6.1b. Proof: Suppose the outer sphere S is earthed and the inner sphere P has a charge $+Q$. An induced charge $-Q$ appears on S. Thus the potential of $P = +Q/4\pi\varepsilon_0 a - Q/4\pi\varepsilon_0 b$; and hence $Q/V = 4\pi\varepsilon_0 ab/(b-a) = C$.

Note that if the *inner* sphere is earthed and the outer sphere charged, there are two capacitors. One is the concentric spherical capacitor of $C = 4\pi\varepsilon_0 ab/(b-a)$. The other is the capacitor formed between the outer sphere and the surrounding earth, of $C = 4\pi\varepsilon_0 b$. The total capacitance is the sum of the two values of C; which simplifies to $4\pi\varepsilon_0 b^2/(b-a)$.

For a *parallel plate capacitor*, $C = \varepsilon A/d$, where ε is the permittivity of the medium, A is the common area between the plates and d is their distance apart. Fig. 6.1c.

Units: C in farad (F) when A in metre^2 and d in metre. $\varepsilon = \varepsilon_r \varepsilon_0$, where $\varepsilon_0 = 8.85 \times 10^{-12}$.

Proof: Electric intensity E between plates $= \sigma/\varepsilon =$ potential gradient $= V/d$, where V is the p.d.

Also, $\sigma = Q/A$, where Q is the charge on either plate.

$$\therefore \quad \frac{Q}{\varepsilon A} = \frac{V}{d}$$

$$\therefore \quad \frac{Q}{V} = \frac{\varepsilon A}{d} = C$$

This is an approximate formula owing to the edge-effect (Fig. 5.4).

EXAMPLE

Find the capacitance of a parallel plate capacitor made from square plates of side 20 cm, with a dielectric of relative permittivity 5 and thickness 2 mm completely filling the space between them.

We have

$$A = 20 \times 20 \text{ cm}^2 = 400 \times 10^{-4} \text{ m}^2, \, d = 2 \text{ mm} = 2 \times 10^{-3} \text{ m}.$$

$$\varepsilon = \varepsilon_r \varepsilon_0 = 5 \times 8.85 \times 10^{-12}$$

$$\therefore \quad C = \frac{\varepsilon A}{d} = \frac{5 \times 8.85 \times 10^{-12} \times 4 \times 10^{-2}}{2 \times 10^{-3}} \text{ F}$$

$$= 9 \times 10^{-10} \text{ F (approx.)}$$

Mixed dielectric. Suppose a parallel plate capacitor has a glass slab of thickness t between the plates, distance d apart, forming a glass dielectric of permittivity ε and an air dielectric of permittivity ε_0. Fig. 6.1d. Then $E_{\text{glass}} = \sigma/\varepsilon$, $E_{\text{air}} = \sigma/\varepsilon_0$. When a charge is taken through both dielectrics from one plate to the other, work done per unit charge $= V$.

$$\therefore \quad E_{\text{glass}} \times t + E_{\text{air}} \times (d-t) = V$$

$$\therefore \quad \frac{\sigma t}{\varepsilon} + \frac{\sigma(d-t)}{\varepsilon_0} = V$$

Substituting $\sigma = Q/A$ and simplifying, we find $C = \varepsilon_0 A/[d - t(1 - 1/\varepsilon_r)]$, where $\varepsilon_r = \varepsilon/\varepsilon_0 =$ dielectric constant of glass.

Demonstration of factors influencing parallel-plate capacitance. Connect the two plates A, B say, to the cap and earthed case respectively of an electroscope. Charge A, and the leaf diverges by an amount which is a measure of the p.d. V between A, B.

(*i*) Move B nearer to A, i.e. decrease *d*. The divergence of the leaf diminishes, so that the p.d. *V* decreases. Thus $C(= Q/V)$ increases, as *Q* is constant.

(*ii*) Move B parallel to A so that the common area diminishes. The divergence of the leaf increases, showing that *V* increases. Thus $C(= Q/V)$ diminishes when the common area diminishes.

(*iii*) Insert well-dried paper between the plates. The divergence decreases, showing that *V* decreases. Hence $C(= Q/V)$ increases with paper as the dielectric in place of air.

Practical capacitors. These are (*i*) *mica capacitor* (a multiplate capacitor with alternate plates connected together), (*ii*) *paper capacitor* (a rolled capacitor with paraffin-waxed paper between the tinfoil strips), (*iii*) *variable capacitor* (a multiplate air capacitor with two sets of plates, one set variable and the other fixed), (*iv*) *electrolytic capacitor* (aluminium plates in ammonium borate solution, with aluminium oxide on one plate to act as the dielectric). The mica and paper capacitors have capacitances from low values to high values such as 4 μF, the variable air capacitor has a maximum capacitance of about 0·0005 μF, and the electrolytic capacitor may have a very high capacitance such as 100 μF or more.

Large and small capacitance. Determination of ε_0.

Large capacitances, of the order of a microfarad, can be compared by a ballistic galvanometer. See Fig. 6.2.

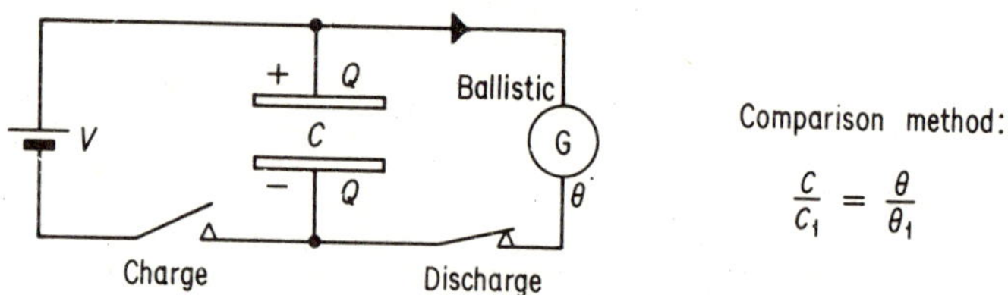

FIG. 6.2 Large capacitance measurement

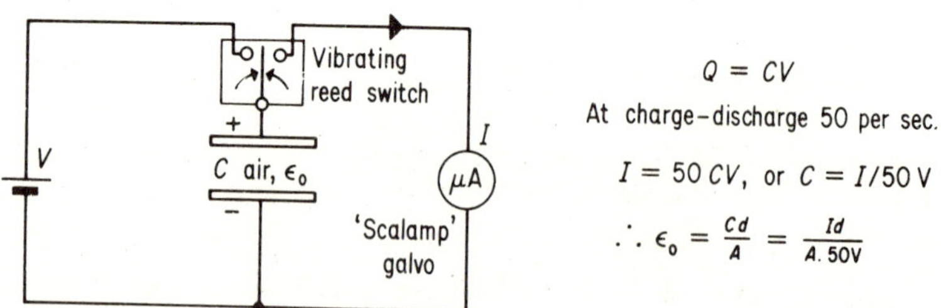

FIG. 6.3 Measurement of small capacitance and of ε_0

Small capacitances, with an air dielectric for example, may be measured by means of a Scalamp galvanometer and vibrating reed switch. Fig. 6.3. Using a d.c. supply V, and charging and discharging C say 50 times per second, then

$$\text{current } I = 50 \, CV.$$

Thus $C = I/50 \, V$. When I is in amperes and V in volts, then C is in farads.

To obtain a value for ε_0, use a large parallel-plate air capacitor of area A and separation d, and determine C as above. Then, since $C = \varepsilon_0 A/d$, $\varepsilon_0 = C.d/A$.

Capacitors in parallel. $\qquad\qquad C = C_1 + C_2.$

The p.d. V across each capacitor is the same; the charge on one capacitor $= C_1 V$, the charge on the other capacitor $= C_2 V$. Fig. 6.4a. The total energy, W, $= \frac{1}{2}(C_1 + C_2)V^2 = \frac{1}{2}CV^2$ (see below)

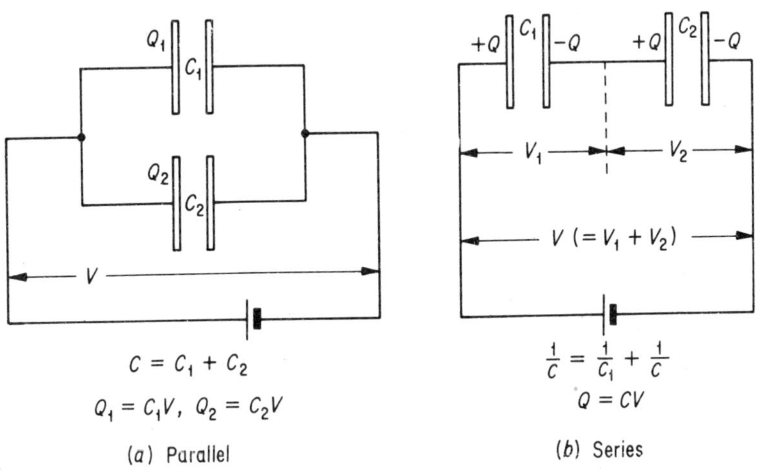

FIG. 6.4 Capacitors in (a) parallel, (b) series

Capacitors in series. $\qquad\qquad \dfrac{1}{C} = \dfrac{1}{C_1} + \dfrac{1}{C_2}.$

The p.d.s., V_1, V_2, across each capacitor are related to the total p.d. V applied by $V = V_1 + V_2$. The charge on each capacitor, Q, $=$ the charge on the equivalent capacitor of capacitance $C = CV$. Fig. 6.4b. The total energy $= Q^2/2C$ (see below).

Energy in capacitors. Energy, $W = \frac{1}{2}CV^2 = \frac{1}{2}QV = \frac{1}{2}\dfrac{Q^2}{C}.$

When Q, C, V are in coulomb, *farad*, volt, then W is in *joule*.

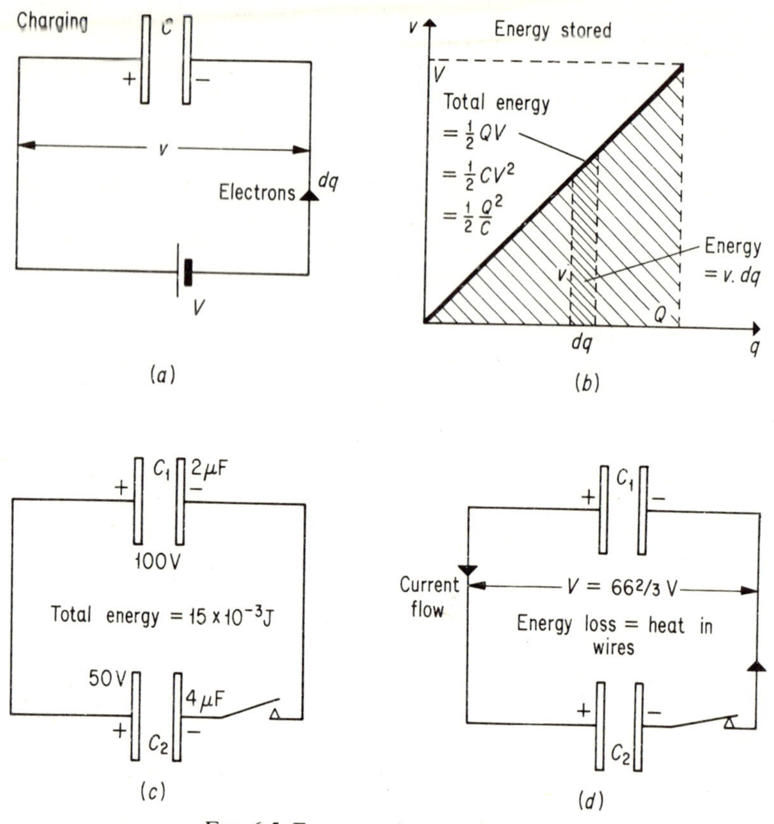

FIG. 6.5 Energy and energy loss.

Proof of energy formula.

Energy, W, = charge × average p.d. between plates

$$= Q \times \frac{V}{2} = \tfrac{1}{2}QV = \tfrac{1}{2}CV^2 = \tfrac{1}{2}\frac{Q^2}{C}.$$

Otherwise, energy, W, $= \displaystyle\int_0^V q.dv = \int_0^V Cv.dv = \tfrac{1}{2}CV^2$

$$= \tfrac{1}{2}QV = \tfrac{1}{2}\frac{Q^2}{C}. \quad \text{Fig. 6.5 } a, b.$$

Energy variations of capacitors. (*a*) Suppose a capacitor of capacitance C is connected to a battery of e.m.f. V and then disconnected from the source. The energy is then given by $\tfrac{1}{2}\dfrac{Q^2}{C}$. When the plates are kept insulated and moved further apart, (*i*) Q

remains constant, (*ii*) C decreases; thus the energy, $Q^2/2C$, *increases*. (*b*) Suppose that the capacitor is connected permanently to a battery of e.m.f. V, and the plates are moved further apart. Then (*i*) the p.d. V remains constant, (*ii*) C decreases; thus, since the energy. W, is $\frac{1}{2}CV^2$, the energy decreases. The charge Q on the plates diminishes also, and hence the capacitor discharges through the battery and restores energy to the battery. When C increases, current is taken from the battery, which therefore loses energy to the capacitor.

Connected charged objects or capacitors. When two charged objects (or capacitors) are joined by a thin wire, (1) *the total charge remains constant*, (2) the potentials of the objects (or the p.d.s. across each capacitor) are equalized, (3) energy is lost as heat in the wire. Problems on connected objects can thus be solved by letting V be the "final" potential (or p.d.) and then applying condition (1).

When two charged capacitors or conductors at a different p.d. are connected together, a current flows. Heat is therefore produced in the connecting wires. The heat energy comes from the store of energy in the capacitors, which is therefore *less* after connection, as the following examples illustrate.

EXAMPLES

1. A 2 μF-capacitor is charged by a p.d. of 100 V; a 4 μF-capacitor is charged by 50 V. If the plates carrying similar charges are connected together by wires, calculate the heat produced in the wires. Fig. 6.5c, d.

Initial total energy $= \frac{1}{2}C_1V_1{}^2 = \frac{1}{2}\times2\times10^{-6}\times100^2 + \frac{1}{2}\times4\times10^{-6}\times50^2$
$= 15\times10^{-3}$ J.

Suppose V is the common p.d. after connection. Then, since total charge is constant,

$$(2\times10^{-6}+4\times10^{-6})\,V = 2\times10^{-6}\times100+4\times10^{-6}\times50$$

$$\therefore \quad V = \frac{200}{3}\,V = 66\tfrac{2}{3}V$$

$$\therefore \quad \text{final total energy} = \frac{1}{2}(2\times10^{-6}+4\times10^{-6})\left(\tfrac{200}{3}\right)^2 = 13\times10^{-3}\,\text{J (approx.)}$$

$$\therefore \quad \text{loss in energy} = \text{heat produced} = 2\times10^{-3}\,\text{J}$$

2. Two spheres of radius 2 and 3 cm are charged with $+16$ and -9×10^{-9} C of charge respectively. They are joined together by a thin metal wire of negligible capacitance. Calculate (*a*) the potential of each sphere, and (*b*) the energy associated with the charges on the spheres before and after joining, in each case.

Before joining. (a) Since $V = Q/C$, and $C = 4\pi\varepsilon_0 r$, where $1/4\pi\varepsilon_0 = 9 \times 10^9$ (approx.),

$$\therefore \quad V_1 = \frac{+16 \times 10^{-9}}{4\pi\varepsilon_0 \times 2 \times 10^{-2}} = \frac{9 \times 10^9 \times 16 \times 10^{-9}}{2 \times 10^{-2}} = 7200 \text{ V}$$

and

$$V_2 = \frac{-9 \times 10^{-9}}{4\pi\varepsilon_0 \times 3 \times 10^{-2}} = \frac{-9 \times 10^9 \times 9 \times 10^{-9}}{3 \times 10^{-2}} = -2700 \text{ V}$$

The energy $W_1 = \tfrac{1}{2}C_1 V_1{}^2 = \tfrac{1}{2} \times 4\pi\varepsilon_0 \times 2 \times 10^{-2} \times 7200^2 = 57 \cdot 6 \times 10^{-6} \text{ J}$

and energy $W_2 = \tfrac{1}{2}C_2 V_2{}^2 = \tfrac{1}{2} \times 4\pi\varepsilon_0 \times 3 \times 10^{-2} \times 2700^2 = 12 \cdot 2 \times 10^{-6} \text{ J}$

After joining. (a) The total charge is constant. Thus if V is the final potential,

$$(C_1 + C_2)V = C_1 V_1 + C_2 V_2 = Q_1 + Q_2$$

$$\therefore \quad (4\pi\varepsilon_0 \times 2 \times 10^{-2} + 4\pi\varepsilon_0 \times 3 \times 10^{-2})\, V = (16 - 9) \times 10^{-9}$$

$$\therefore \quad V = \frac{7 \times 10^{-9}}{4\pi\varepsilon_0 \times 5 \times 10^{-2}} = 1260 \text{ V}$$

$$\therefore \quad \text{energy of 1}^{st} \text{ sphere} = \tfrac{1}{2}C_1 V^2 = \tfrac{1}{2} \times 4\pi\varepsilon_0 \times 2 \times 10^{-2} \times 1260^2$$

$$= 1 \cdot 8 \times 10^{-6} \text{ J}$$

and $\quad$ energy of 2nd sphere $= \tfrac{1}{2}C_2 V^2 = \tfrac{1}{2} \times 4\pi\varepsilon_0 \times 3 \times 10^{-2} \times 1260^2$

$$= 2 \cdot 7 \times 10^{-6} \text{ J.}$$

Some electrometer measurements. The UNILAB apparatus has a form of electrometer triode with a variable high input resistance R and high input capacitance C, followed by a transistor amplifier for amplifying d.c. current. A $0 - 1$mA meter is connected to the output terminals. If $R = 100$k megohms $= 10^{11}$ ohms, and a full scale deflection of the meter is obtained when 1 V is applied, the current flowing is then 10^{-11}A.

The following measurements may be made:

(1) *Ionisation currents.* Suppose two metal parallel plates are joined to the input terminals and a high p.d. is applied between them. If the air between the plates is ionised, the meter reading gives a direct value of the ionisation current.

(2) *Charge and small capacitance.* With the input switch at C only, the input capacitor is 0·001 μF. A full scale meter reading with 1 V applied then corresponds to a charge $Q = CV = 0·001 \times 10^{-6} \times 1 = 10^{-9}$C. With an input capacitor of 0·01 μF the meter is less sensitive, as a full scale reading then indicates 10^{-8}C.

When the capacitance of a charged object is small compared to the internal capacitor, the charge can be measured directly by connecting it to the capacitor, as practically the whole of it is then transferred. Suppose a metal sphere X of radius 5 cm, capacitance C, is charged and the quantity Q is read by the meter. On sharing the charge Q with a small capacitor C_1 such as a gold leaf electroscope Y, the reduced charge Q_1 on the sphere can be measured in the same way. Then, since the potential of X and Y are the same after connection,

$$\frac{Q_1}{C_1} = \frac{Q-Q_1}{C}, \qquad \text{or} \quad C_1 = C\left(\frac{Q_1}{Q-Q_1}\right)$$

Knowing $C = 4\pi\varepsilon_0 r$ and measuring Q_1 and Q, then C_1 can be calculated.

(3) *Dielectric constant.* This can be found by measuring the charge on two parallel metal plates before and after the dielectric is inserted to completely fill the space between them. Alternatively, use the apparatus in Fig. 6.3 and compare the currents before and after the dielectric is used.

Charge and discharge through high resistance.*Charging.* The charge Q grows according to an exponential law to its final charge Q_0. Formula : $Q = Q_0(1 - e^{-t/CR})$, where C is in farads, R is in ohms and t is in seconds. The *time constant* $= CR$ seconds $=$ the time for the charge to reach 63 per cent (approx.) of its final charge.

Discharging. Q diminishes from its original charge Q_0 according to $Q = Q_0 e^{-t/CR}$; the time constant $= CR$. $5\,CR$ may be taken roughly as the time for the capacitor to discharge completely. If the capacitor requires intermittent discharge, a neon tube can be placed across it in the circuit (simple time-base circuit).

Proof : Diminishing current $I = -\dfrac{dQ}{dt} = \dfrac{V}{R} = \dfrac{Q}{CR}$

$$\therefore \int_{Q_0}^{Q} \frac{dQ}{Q} = \int_0^t -\frac{dt}{CR}, \quad \text{or} \quad Q = Q_0 e^{-t/CR}$$

Discharge of capacitor through coil. *Oscillations* of current and charge are obtained (compare the discharge through a resistance, when the current and charge diminish exponentially). There is an interchange of energy between the capacitor (energy in electrostatic field) and the coil (energy in magnetic field). See Fig. 2.4, p. 17.

Proof : For a coil of inductance L and negligible resistance, $V = -L dI/dt$ (p. 99).

$$\therefore \quad -L\frac{d^2Q}{dt^2} = \frac{Q}{C}, \quad \text{or} \quad \frac{d^2Q}{dt^2} = -\frac{1}{LC} \cdot Q.$$

$$\therefore \quad \text{period of oscillation, } T, = 2\pi\sqrt{LC}, \text{ or frequency}, f, = \frac{1}{2\pi\sqrt{LC}}.$$

With a coil of inductance and resistance, *damped* oscillations are obtained when a capacitor is discharged through it; energy is lost as heat in the resistance.

Current Electricity

7. CONDUCTORS. CIRCUITS

Charge carriers. In *metals*, current is carried by electrons ($I = nevA$, p. 121); in *electrolytes*, by ions; in *semiconductors*, by electrons and holes (equivalent to charges $+e$). Fig. 7.1a, b, c.

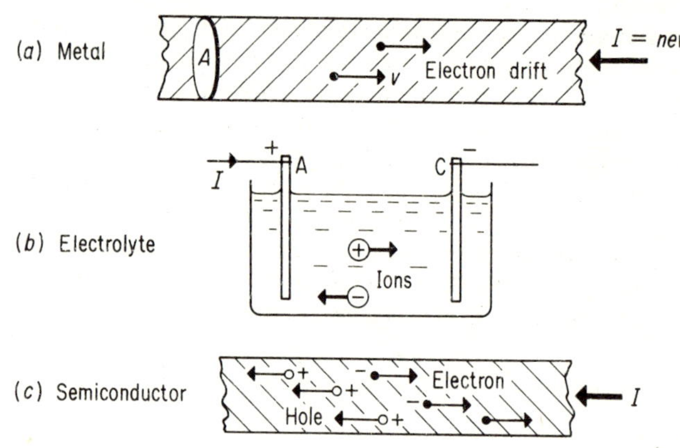

FIG. 7.1 Charge carriers.

Formulae and units. (1) For a resistor, $I = V/R$, $R = V/I$, $V = IR$. I is in ampres (A), V in volts (V), R in ohms (Ω).

1 milliampere (mA) = 1/1000 A, 1 microhm = $10^{-6}\,\Omega$, 1 megohm ($M\Omega$) = $10^6\,\Omega$.

$$(2) \qquad R = \rho \frac{l}{A},$$

where ρ is the *resistivity*, l is the length of wire, A its cross-sectional area. If the diameter of a wire is 0·56 mm, then $A = \pi r^2 = \pi \times 0\cdot28^2 \times 10^{-6}\,\text{m}^2$. SI unit of ρ: ohm metre (Ω m).

$$(3) \qquad R_t = R_0(1 + \alpha t),$$

where α is the *mean temperature coefficient* of resistance, R_0 is the resistance at 0°C, and t is the temperature in °C (see also p. 57). The unit of α is "K^{-1} (degC^{-1})". If the resistance of a wire at 16°C is 6·42 Ω and at 80°C it is 7·35 Ω, then

$$7{\cdot}35 = R_0(1+80\alpha), \qquad 6{\cdot}42 = R_0(1+16\alpha).$$

$$\therefore \quad \frac{7{\cdot}35}{6{\cdot}42} = \frac{1+80\alpha}{1+16\alpha}.$$

(4) *Resistance in series.* Total resistance $R = R_1 + R_2$.

(i) The current is the same in each wire.

(ii) The p.d. across each wire is proportional to its resistance. Thus if a p.d. of 2V is applied across wires of 95Ω and 5Ω in series, the p.d. across the 5Ω wire $= \dfrac{5}{100} \times 2V = 0{\cdot}1V$, and the p.d. across the 95-ohm wire $= \dfrac{95}{100} \times 2V = 1{\cdot}9V$.

(5) *Resistances in parallel.* Total resistance R is given by

$$\frac{1}{R} = \frac{1}{R_1} + \frac{1}{R_2}.$$

(i) The p.d. across each wire is the same.

(ii) The current in each wire is inversely proportional to its resistance. Thus if a current of 4A flows towards the junction of two wires of 80 and 20 Ω in parallel, the currents in the 80 Ω and 20 Ω wire are in the proportion $\dfrac{1}{80} : \dfrac{1}{20}$, or 20 : 80; i.e. the current in the 80 Ω wire $= \dfrac{20}{20+80} \times 4 = 0{\cdot}8$ A. With three wires of 5, 10, 20 Ω in parallel, the current is shared in the ratio $\dfrac{1}{5} : \dfrac{1}{10} : \dfrac{1}{20}$, or 4 : 2 : 1.

The p.d. across a parallel combination of wires can be calculated from $V = IR$, where R is the combined (total) resistance and I is the current flowing in the main circuit, outside the parallel arrangement.

EXAMPLES

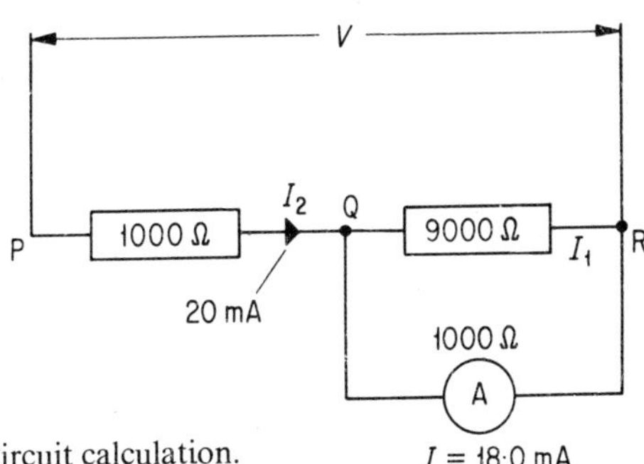

Fig. 7.2 Circuit calculation.

1. In Fig. 7.2, the current in the meter A of 1000 Ω resistance is 18 mA. Calculate the p.d. V applied across the whole circuit.

The current I_1 in the branch QR is 9 times as *small* as 18 mA, since the resistance in QR is 9 times as large as 1000 Ω.

$$\therefore \quad I_1 = 2 \text{ mA}.$$

$$\therefore \quad \text{current } I_2 \text{ in PQ} = 2 \text{ mA} + 18 \text{ mA} = 20 \text{ mA}$$

$$\therefore \quad \text{p.d. across PQ} = I_2 R = \frac{20}{1000} \times 1000 = 20 \text{ V}$$

$$\text{and} \quad \text{p.d. across QR} = I_1 \times 9000 = \frac{2}{1000} \times 9000 = 18 \text{ V}$$

$$\therefore \quad \text{total p.d. } V = 20 \text{ V} + 18 \text{ V} = 38 \text{ V}.$$

2. The temperature coefficient of copper is 4×10^{-3} K^{-1} ($^\circ$C^{-1}). Find the resistance of 5 metres of copper wire, diameter 0·2 mm, at 100°C if the resistivity at 0°C is $1 \cdot 7 \times 10^{-8}$ Ω m.

$$\text{At } 0^\circ\text{C, resistance } R_0 \text{ of wire} = \rho \frac{l}{A} = \frac{1 \cdot 7 \times 10^{-8} \times 5}{\pi \times (0 \cdot 1 \times 10^{-3})^2}$$

$$= 2 \cdot 7 \ \Omega$$

$$\therefore \quad \text{resistance } R_{100} \text{ at } 100^\circ\text{C} = R_0 (1 + \alpha t)$$

$$= 2 \cdot 7 \ (1 + 4 \times 10^{-3} \times 100) = 3 \cdot 8 \ \Omega.$$

Measurement of p.d. with moving-coil voltmeter. When p.d. is measured with a moving-coil voltmeter, the resistance of the voltmeter is effectively *in parallel* with the resistance S across which it is placed. The current in the circuit alters when the voltmeter is used, as the total resistance in the circuit changes, and thus the p.d. across S is altered. To calculate the new p.d. use $V = IR$, where I is the new current in the circuit and R is the combined resistance of S and the voltmeter in parallel. The resistance of a moving-coil voltmeter should be at least ten times the resistance across which it is placed in order to measure the p.d. accurately.

EXAMPLE

Two resistors, A and B, are 60,000 and 40,000 Ω respectively and are in series across a constant supply of 100 V. A 10,000 Ω voltmeter is connected across B. What is the p.d. across B (*i*) before (*ii*) after the voltmeter is connected?

(*i*) *Before* the voltmeter is connected,

$$\text{p.d. across B} = \frac{40,000}{60,000 + 40,000} \times 100 \text{ V} = 40 \text{ V}$$

(*ii*) *After* the voltmeter is connected, total resistance R across B is given by

$$\frac{1}{R} = \frac{1}{40,000} + \frac{1}{10,000} = \frac{5}{40,000}$$

$$\therefore \quad R = 8,000 \; \Omega$$

$$\therefore \quad \text{new p.d. across B} = \frac{8,000}{60,000 + 8,000} \times 100 \text{ V}$$

$$= 12 \text{ V (approx.)}$$

The considerable drop of p.d. from 40 V to 12 V is due to the relatively low resistance of the voltmeter compared to the resistance, 40,000 Ω, across which it is joined.

Ohm's law. This states: Provided the physical conditions, such as temperature, do not alter, the ratio V/I is a constant for a given conductor where V, I are the p.d. and current respectively.

In proving this law by an experiment we must not use a moving-coil voltmeter to measure V, as the instrument has been calibrated assuming that Ohm's law is true. We can use a potentiometer for measuring V (see p. 65) and a milliammeter for measuring I, together with a resistor of the order of 1000 ohms. The $I-V$ graph obtained is a straight line passing through the origin. Fig. 7.3*a*.

Many substances do not obey Ohm's law, as discussed on p. 58.

Variation of resistance with temperature. The resistance of a pure metal increases with increasing temperature (positive temperature coefficient); the resistance of carbon and non-metals decrease with increasing temperature (negative temperature coefficient); the resistance of eureka, manganin and nichrome (alloys) change very slightly as the temperature changes (extremely low temperature coefficient).

Over a *wide* range of temperature, the resistance R_t of a pure metal is related to its resistance R_0 at $0°C$ by

$$R_t = R_0(1 + \alpha t + \beta t^2),$$

where t is the temperature on the gas thermometer (standard) Celsius scale, and β, α are constants; β is much smaller than α, which is about $0 \cdot 004 \text{ K}^{-1}$ (deg C^{-1}).

Over a *limited* range of temperature, R_t may be represented by the equation $R_t = R_0(1 + \alpha t)$; in this case α is called the "mean temperature coefficient" in the range of temperature concerned.

Characteristic curves. Rectifiers. The "characteristic" of a substance is the current-p.d. (I-V) curve obtained for it. A substance obeying Ohm's law, such as a *pure metal*, has a straight line characteristic passing through the origin; the line is continuous when the p.d. is reversed, i.e. when $-V$ and $-I$ are plotted, showing the resistance remains the same. Fig. 7.3a.

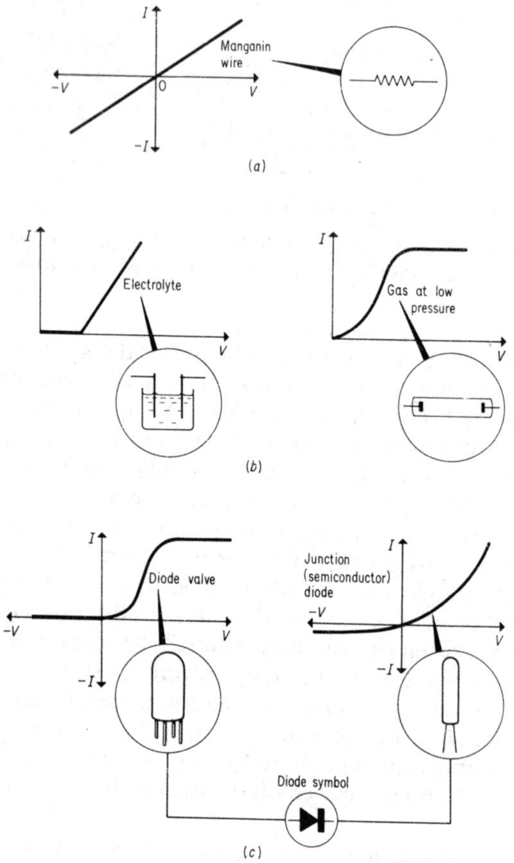

FIG. 7.3 Ohmic and non-ohmic conductors.

When a p.d. is applied to platinum electrodes dipping into dilute sulphuric acid, no current flows until the applied p.d. just exceeds the back e.m.f. due to the hydrogen and oxygen formed (about 1 volt). After this the I-V curve is a straight line. The arrangement does not obey Ohm's law as the straight line does not pass through the origin. Fig. 7.3b. The I-V characteristic of a gas does not obey Ohm's law.

A copper disc oxidized on one face, forming a cuprous oxide-copper layer, has a *non-linear characteristic*. In one direction of applied p.d. the characteristic is a curve; in the other direction only a small current flows as V is increased. The disc thus has a high resistance in one direction and a low resistance in the reverse direction. It is called a *rectifier*, and as it does not obey Ohm's law it is called a "non-ohmic conductor".

A *diode valve* characteristic has a small initial curvature followed by a straight line, and then the curve flattens (saturation current now obtained). When the p.d. is reversed, no current flows. Fig. 7.3c. Thus the valve has a low resistance in one direction, and an infinitely large resistance in the other direction. The valve does not therefore obey Ohm's law. The *junction* (semiconductor) *diode* is also non-ohmic. Fig. 7.3c.

E.M.F. AND INTERNAL RESISTANCE

Electromotive force (e.m.f.). This is the "total energy per coulomb" or "total power per ampere" supplied by the generator or cell. Generally, EI is the total power supplied, VI is the power consumed in the appliance connected where V is the terminal p.d., and vI or I^2r is the power consumed in the internal resistance. Fig. 7.4a.

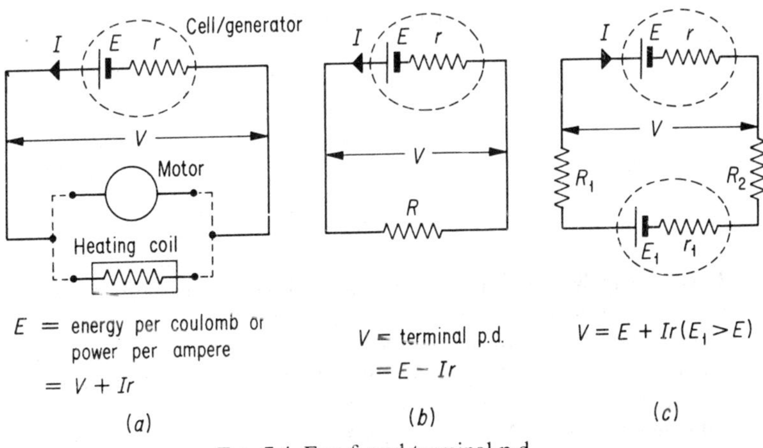

FIG. 7.4 E.m.f. and terminal p.d.

Suppose a cell is joined to a resistance R. Fig. 7.4b. Then, from $EI = VI + I^2r$,

$$\text{terminal p.d. } V = E - Ir \quad . \quad . \quad . \quad . \quad . \quad . \quad (i)$$

and

$$r = \frac{E - V}{I} \quad . \quad . \quad . \quad . \quad . \quad . \quad (ii)$$

Terminal p.d. when current is forced through a cell *in opposition to its e.m.f.*, as in charging an accumulator, is given by

$$V = E + Ir,$$

that is, the terminal p.d. is greater than the e.m.f. Fig. 7.4(c).

Cells in series. The total e.m.f., E, = the algebraic sum of the e.m.fs. of the individual cells. The total internal resistance, r, = the sum of the individual internal resistances. Thus if two cells of e.m.fs 1·5 and 1·1V respectively are *in opposition*, and their internal resistances are 2 and 3 Ω respectively, then

$$\text{total e.m.f.} = 1{\cdot}5 - 1{\cdot}1 = 0{\cdot}4\text{V},$$

$$\text{total internal resistance} = 2 + 3 = 5\Omega.$$

EXAMPLE

In the circuit of Fig. 7.4c, $E_1 = 12\,\text{V}, r_1 = 2\Omega; E = 4\,\text{V}, r = 3\Omega, R_1 = 8\Omega, R_2 = 7\,\Omega$. Find the terminal p.d. across each cell.
We have

$$\text{net e.m.f. } E = 12 - 4 = 8\text{ V (cells in opposition)},$$

$$\text{total resistance } R = 8 + 7 + 2 + 3 = 20\ \Omega$$

$$\therefore\quad \text{current } I = \frac{E}{R} = \frac{8}{20} = 0{\cdot}4\text{ A}$$

$$\therefore\quad \text{terminal p.d. across 12 V cell} = 12 - Ir_1 = 12 - 0{\cdot}4 \times 2 = 11{\cdot}2\text{ V}$$

and $\quad$ terminal p.d. across 4 V cell $= 4 + Ir = 4 + 0{\cdot}4 \times 3 = 5{\cdot}2\text{ V}$

Cells in parallel. When the cells are different, Kirchhoff's laws (see p. 61) must be applied to calculate the current.

When the cells are similar, the total e.m.f. = the e.m.f. of *one* cell, E. The internal resistances are in parallel; thus the combined internal resistance of n cells each of internal resistance $r = r/n$.

Maximum power. A cell delivers maximum power to an external resistance R when $R = r$, the internal resistance. The current then flowing is thus $E/(r + r)$, or $E/2r$.

Proof. Power, P, $= I^2R = E^2R/(R + r)^2$, substituting for I. By differentiating,

$$\frac{dP}{dR} = E^2\left[\frac{r^2 - R^2}{(R + r)^4}\right].$$

$$\therefore\quad \frac{dP}{dR} = 0 \text{ when } R = r,$$

and this can be shown on testing to be a maximum value.

Kirchhoff's Laws. For complicated circuits, Kirchhoff's laws are applied to calculate the currents in the various branches.

Law 1. The algebraic sum of the currents at a junction of an electrical circuit is zero. Thus if a current of x amps flows towards a junction of three branches, and y amps flows away from the junction in another branch, $(x - y)$ amps flows away in the third branch of the junction.

Law 2. The algebraic sum of the e.m.fs. in a closed electrical circuit is equal to the algebraic sum of all the IR products in that circuit.

RESISTANCE MEASUREMENT

Wheatstone bridge. This contains four resistances, P, Q, R, S, forming a closed network. A galvanometer G is placed across one pair of junctions; a battery is placed across the other pair. At a balance, $P/Q = S/R$ (Fig. 7.5).

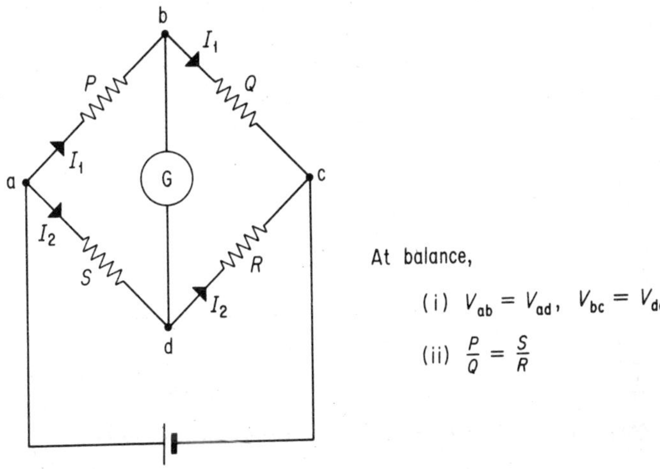

FIG. 7.5 Wheatstone bridge circuit.

The proof of the ratio relation follows from: (*i*) the current I_1 in P, Q is the same at a balance, and the current, I_2, in S, R is the same. (*ii*) The p.d. across $P = I_1 P =$ the p.d. across $S = I_2 S$; the p.d. across $Q = I_1 Q =$ the p.d. across $R = I_2 R$.

The bridge is insensitive for very high resistances, above a few thousand ohms, and for very low resistances, less than 0·1 ohm. For a *very high resistance* R of the order of a megohm, a mirror galvanometer, a small p.d. V, and R may be placed in series, and the current I measured. Then $R = V/I$, neglecting the galvanometer resistance (see also p. 63). A *low resistance* can be measured by comparing it with a standard low resistance by a potentiometer experiment (p. 66).

Metre bridge. The unknown resistance X is placed in one gap, a known resistance R, of the same order, in the other gap. At a balance, $X/R = l_1/l_2$, where l_1, l_2 are the balance-lengths. For accuracy the balance-point must be near the middle of the wire so that the effect of the unknown "end-correction" is small, and X, R are interchanged to obtain another reading. A series high resistor is used with the sensitive galvanometer, as a protective resistor, until a near-balance is obtained.

Post Office box. This has (*i*) ratio arms, 10, 100, 1000Ω, (*ii*) a variable resistance 1–11,000Ω, (*iii*) tapping-keys for the galvanometer and the battery. Before connecting the circuit a theoretical Wheatstone bridge should be drawn on paper, the junctions of the wires labelled a, b, c, d, and the corresponding points then identified on the box. The protective resistor r must be removed near the balance-point (Fig. 7.6).

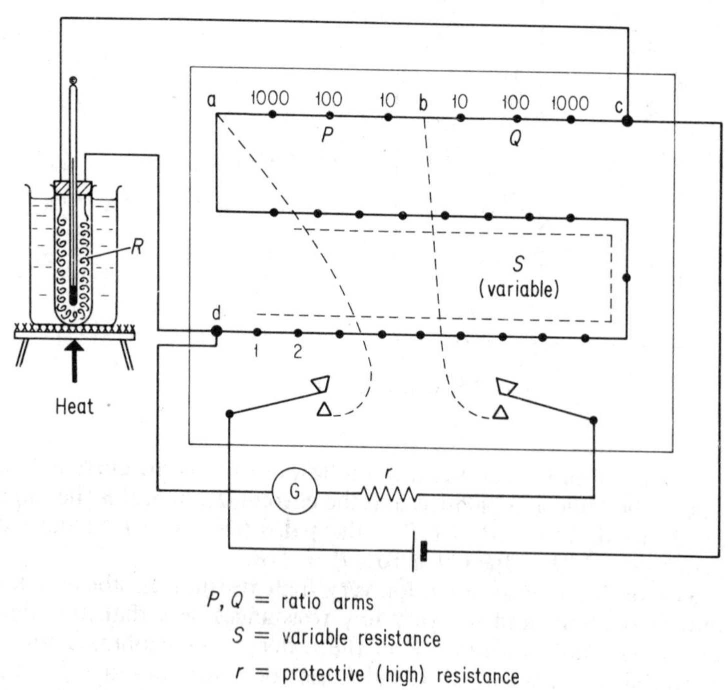

FIG. 7.6 Accurate measurement of resistance—Post Office box.

If P, Q are the ratio arms, R is the unknown resistance, and S is the known resistance, then, at a balance, $R = S \times \dfrac{Q}{P}$. Thus for decimal places $Q = 10$, $P = 100$, or $Q = 10$, $P = 1000$.

Resistivity. The resistance R ohms is measured, the length l in metres is determined from a metre rule, and the diameter d in metres is found with a micrometer screw-gauge. Then resistivity

$$\rho = \frac{RA}{l} = \frac{R . \pi d^2}{4l} \ \Omega \ \text{m}$$

Temperature coefficient. The resistance R_t of the metal is measured at varying temperatures, t°C (Fig. 7.6). R_t v. t is plotted, origin 0°C, and the straight line through the points is produced to cut the R_t axis at some value R_0, which is the resistance at 0°C. Then, using the graph to find a resistance R_t at t°C, the mean temperature coefficient, α, is given by

$$\alpha = \frac{R_t - R_0}{R_0 \times t}.$$

Resistance of galvanometer. The galvanometer is placed in the fourth arm of the bridge, and a small p.d. is placed across one pair of junctions. A tapping-key is placed across the other pair of junctions. The variable resistance on the box is altered until no change in the galvanometer's deflection occurs on pressing the key. A balance is then obtained, and the usual ratio relation holds.

High resistance. The resistance R is connected to a battery and a mirror galvanometer, and the deflection θ is observed. A known high resistance R_1 then replaces R, and a deflection θ_1 is obtained. Then $R/R_1 = \theta_1/\theta$, from which R is calculated. Alternatively, the discharge of a capacitor through the resistance can be used to find R (p. 53).

Electrolyte resistance. An a.c. supply of $f = 1000$ Hz is used instead of a d.c. source. A telephone earpiece is used as the detector in place of the galvanometer. The electrolyte is contained in a capillary tube X joined to two vessels containing the electrolyte, and is placed in one gap of a metre bridge, the other containing a known resistance. At a balance-point there is a minimum sound in the earpiece, and the usual ratio relation gives the electrolyte resistance R_1. A length l of X is then cut off, and the resistance R_2 is obtained again. Then, if ρ is the electrolyte resistivity and A the cross-sectional area of X, $(R_1 - R_2) = \rho l/A$. The *conductivity* $= 1/\rho$.

8. POTENTIOMETER MEASUREMENTS

Comparison of e.m.fs. The p.d. along the wire of the potentiometer is proportional to the length, if the current is constant. When the e.m.f. E_1 of a cell is balanced against the p.d. V_1 along a length l_1 of the wire, and the e.m.f. E_2 of another cell is balanced against a p.d. V_2 along a length l_2, then, Fig. 8.1.,

$$\frac{E_1}{E_2} = \frac{V_1}{V_2} = \frac{l_1}{l_2}.$$

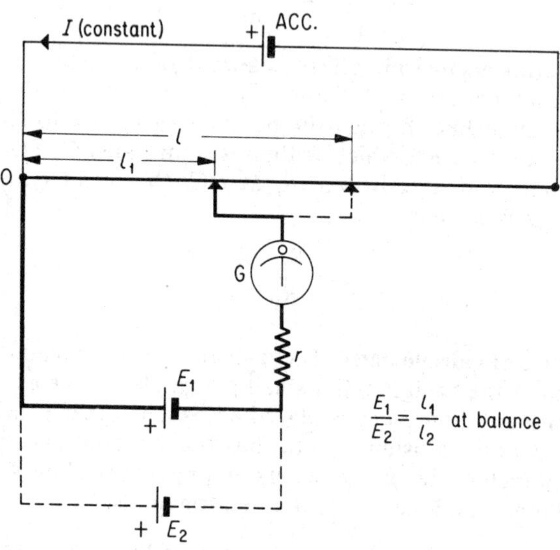

$$\frac{E_1}{E_2} = \frac{l_1}{l_2} \text{ at balance}$$

FIG. 8.1 Potentiometer theory.

To find the actual e.m.f., a *standard cell* is used. This has an e.m.f. E known accurately, for example 1·0186 V. Then, if a balance is obtained 51·4 cm along the wire, $E_1/1·0186 = l_1/51·4$, from which E_1 can be calculated. Also,

$$\text{p.d. per centimetre length of wire} = \frac{1·0186}{51·4} = 0·0197 \text{ V.}$$

Note. (1) In using the potentiometer the positive pole of the cell examined must always be connected to the same end of the wire as the positive pole of the accumulator. (2) An *accumulator* is always used with the wire to obtain a constant current. (3) The p.d. across the wire can be reduced by placing a resistance in series with it and the accumulator. (4) Always use a protective resistance with the galvanometer and remove it to find the final balance. (5) Always re-check the balance-point at the end of the experiment with a standard cell, to see if the accumulator has run down.

Calibration of voltmeter. A potential divider is used to obtain varying readings, on the voltmeter (Fig. 8.2). At each reading the voltmeter terminals are joined to the potentiometer wire and the corresponding balance-length found. The length is then changed to volts from a knowledge of the p.d. per centimetre of the wire, previously found by using a standard cell.

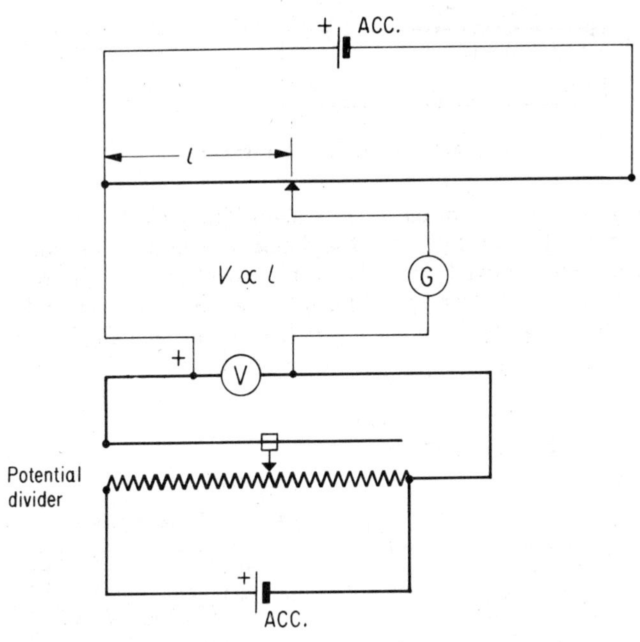

FIG. 8.2 Calibration of voltmeter.

Calibration of ammeter. The ammeter is placed in series with a battery, rheostat and a *standard* 1 Ω resistor R (Fig. 8.3). The p.d., across R is measured on the potentiometer wire, and converted to volts. The current $I = V/R = V/1$, and hence I can be calculated.

Comparison of resistances. A low resistance, less than 0·1 ohm, cannot be measured by the Wheatstone bridge method, as the bridge is then too insensitive. A potentiometer can be used, however. The resistance R_1 is placed in series with a known low resistance R_2,

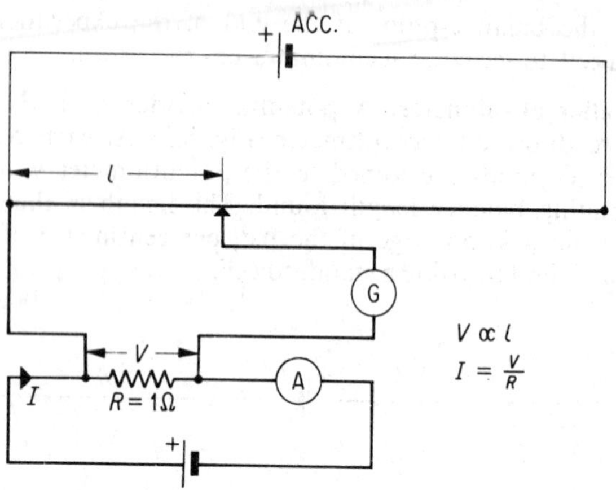

Fig. 8.3 Calibration of ammeter.

a rheostat and an accumulator (Fig. 8.4a). The p.d. V_1 across R_1 is measured on the bridge, and the balance-length l_1 noted. The wires are then removed from the terminals of R_1 and the p.d. V_2 across R_2 is found, corresponding to a length l_2 say. Then $V_1/V_2 = IR_1/IR_2 = R_1/R_2$. Thus $R_1/R_2 = l_1/l_2$, or $R_1 = l_1 R_2/l_2$.

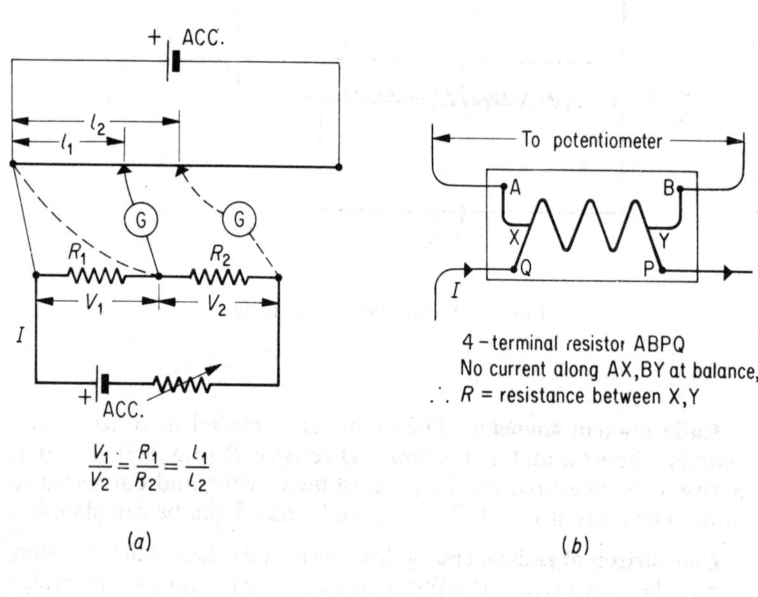

$$\frac{V_1}{V_2} = \frac{R_1}{R_2} = \frac{l_1}{l_2}$$

(a)

(b)

Fig. 8.4 Measurement of low resistance.

For very low resistance, a four-terminal standard resistance is used. Fig. 8.4b. This eliminates error due to contact resistances at the terminals.

Internal resistance of cell. The internal resistance, $r, = (E - V)/I$, p. 59 But $I = V/R$.

$$\therefore \quad r = \left(\frac{E}{V} - 1\right) R.$$

The ratio E/V is found on the potentiometer wire by (*i*) measuring the terminal p.d. E when the cell is on open circuit, (*ii*) measuring the terminal p.d. V when a resistance R is connected (Fig. 8.5).

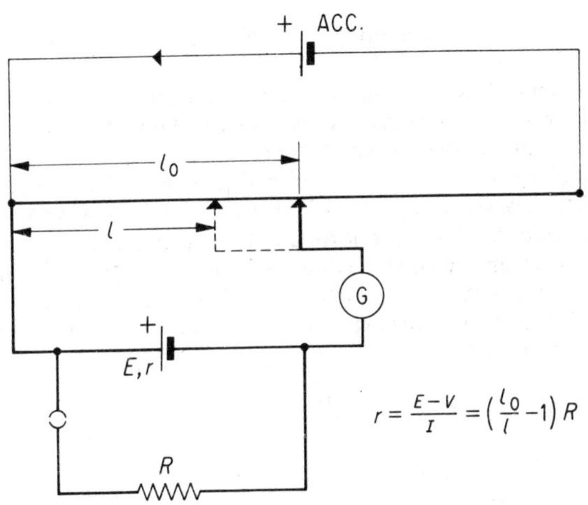

FIG. 8.5 Internal resistance of cell.

If a graph is required, $1/R$ is plotted against $1/V$. Then, since $r/R = E/V - 1$, we have $r = -R$ when $1/V = 0$; so that r can be found from the intercept of the straight line graph on the $1/R$ axis.

EXAMPLES

1. On open circuit, the terminal p.d. of a cell is balanced by 75·0 cm of potentiometer wire. When a 4 Ω resistance is joined to the terminals, this p.d. is now balanced by 60·0 cm of wire. Calculate the internal resistance of the cell.

We have

$$\frac{E}{V} = \frac{75 \cdot 0}{60 \cdot 0} = \frac{5}{4}$$

$$\therefore \quad r = \left(\frac{E}{V} - 1\right) R = \left(\frac{5}{4} - 1\right) 4 = 1 \, \Omega.$$

2. An ammeter is in series with a standard 1-ohm resistor and the current flowing is registered on the meter as 1·20 A. The p.d. across the register is balanced by a length of 64·0 cm of a potentiometer wire and the e.m.f. or 1·02 V of a standard cell is balanced by a length of 50·6 cm of the same wire. Calculate the error in the ammeter reading.

Suppose V is the p.d. across the 1-ohm. Then

$$\frac{V}{1\cdot02} = \frac{64\cdot0}{50\cdot6}, \quad \text{or} \quad V = 1\cdot29 \text{ V}$$

$$\therefore \quad \text{current } I = \frac{V}{R} = \frac{1\cdot29}{1} = 1\cdot29 \text{ A}$$

$$\therefore \quad \text{error in ammeter} = 1\cdot20 - 1\cdot29 = -0\cdot09 \text{ A}.$$

3. Give a labelled diagram showing how a potentiometer may be used to compare the resistance of two resistors. Describe how the experiment would be carried out and give the necessary theory.

An accumulator of e.m.f. 2·10 V and with negligible internal resistance is connected in series with a resistor of 5·00 Ω and a uniform slide wire of length 200 cm and resistance 10·0 Ω to form a potentiometer. A cell of e.m.f. 1·08 V and internal resistance 6·00 Ω is connected with the appropriate polarity to the slide wire, through a galvanometer. Find the balancing length. Discuss quantitatively the effect on the experiment of connecting a 1·00 ohm resistor in parallel with the cell. (L.).

(1) $\qquad$ P.d. across slide wire $= \dfrac{10}{10+5} \times 2\cdot1 \text{ V} = 1\cdot4 \text{ V}$

At balance, p.d. = e.m.f. = 1·08 V

$$\therefore \quad \text{balance length} = \frac{1\cdot08}{1\cdot4} \times 200 \text{ cm} = 154 \text{ cm (approx.)}$$

(2) With a 1·00 ohm in parallel with the cell, the terminal p.d. V = p.d.

across 1·00 ohm $= IR = \left(\dfrac{1\cdot08}{6+1}\right) \times 1 = 0\cdot154 \text{ V}$

$$\therefore \quad \text{new balance length} = \frac{0\cdot154 \text{ V}}{1\cdot4 \text{ V}} \times 200 \text{ cm} = 22 \text{ cm}$$

Measurement of thermo-electric e.m.f. A small p.d. of the order of a millivolt can be measured by *placing a resistance R in series with the potentiometer wire* (Fig. 8.6). Thus suppose the wire has a resistance of 4 Ω, and the accumulator an e.m.f. of 2 V. The current, $I, = 2/(R+4)$. Hence the p.d. V across the wire, 1 millivolt say, is given by IR, or $2 \times 4/(R+4)$.

$$\therefore \frac{8}{R+4} = \frac{1}{1000}$$

$$R = 7996\ \Omega.$$

Thus R must be 7996 Ω to obtain a p.d. of 1 millivolt (1000 microvolts) across the 100-cm wire. A p.d. of 50 microvolts will then be balanced at a length $50 \times 100/1000$ or 5 cm.

Fig. 8.6 illustrates the measurement of a thermoelectric e.m.f. E_1 between two unlike metals A, B, such as copper and constantan, whose junctions are at different temperatures.

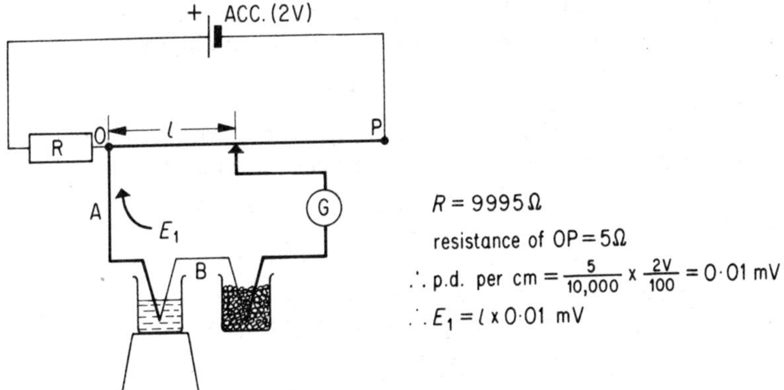

FIG. 8.6 Thermoelectric e.m.f. measurement.

For accurate measurement, R is obtained from two post office boxes in series. A standard cell is placed across one box, and resistances are taken out of one box and put back in the other, until a balance is obtained with the standard cell; during the interchange, the *total* resistance R is kept constant. The p.d. per ohm of the circuit is thus accurately found, and the p.d. across the wire is hence known accurately.

Measurement of large p.d. and current by potentiometer. A *large p.d.* such as 100 V can be measured by applying the p.d. across a number of resistors in series, so that a p.d. smaller than 2 V is obtained across one of them, resistance r say. The p.d. v across r is measured by the potentiometer. Then if V is the large p.d., and R is the total resistance of the series resistors, $V = vR/r$.

A *large current* such as 100 A can be measured by passing it through a low resistance of 0·01 Ω for example, made of eureka or manganin, and measuring the p.d. V by the potentiometer. The current is then $V/0·01$ or $100V$.

9. ELECTRICAL ENERGY. ELECTROLYSIS

ELECTRICAL ENERGY

Electric formula, and heating effect. In deriving the formula for electrical energy, or the heat produced by a current, begin with *the definition of unit p.d., the volt.*

One volt is defined as the p.d. between two points when 1 joule of work is done or energy released when 1 coulomb is taken between the two points.

When a charge Q is transferred between two points having a p.d. V, the energy W released is thus given by

$$W = QV.$$

W is in joules when Q is in coulombs and V in volts.

If the current is I and the time t, $Q = It$; and hence

$$W = IVt.$$

When a "passive" resistance R is used, the whole of the energy supplied is converted to *heat*. In this case, since $I = V/R$,

$$W = I^2Rt \text{ or } V^2t/R \text{ (Fig. 9.1}a).$$

(*a*) Resistor

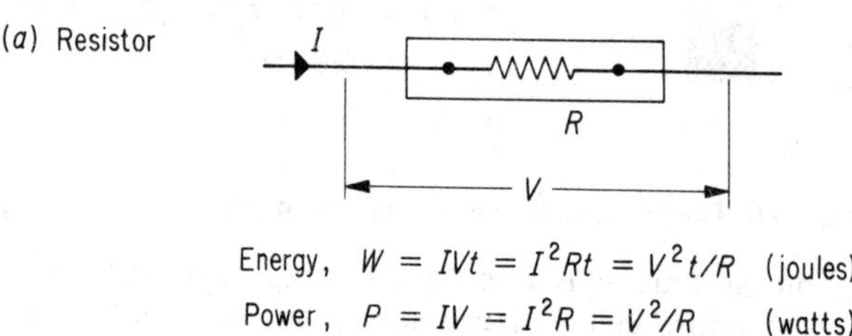

Energy, $W = IVt = I^2Rt = V^2t/R$ (joules)

Power, $P = IV = I^2R = V^2/R$ (watts)

(*b*) High tension transmission

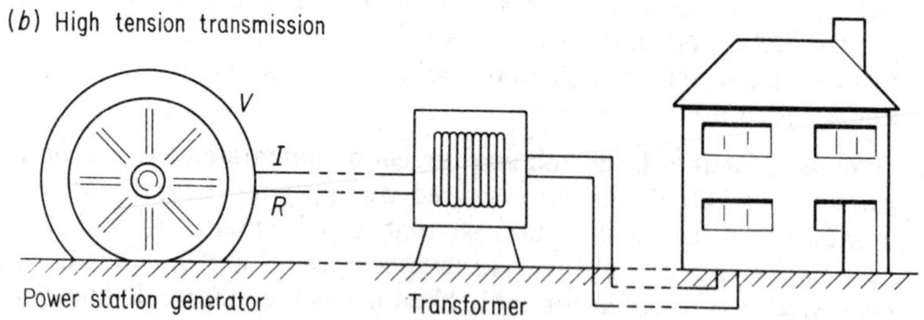

High voltage, small current I;
low power losses (I^2R), thin cable

FIG. 9.1 Electrical energy.

Electrical power, P, = energy per second = $\dfrac{\text{energy, } W}{\text{time, } t}$

$$\therefore \quad P = IV = I^2R = \frac{V^2}{R}$$

When I is amperes, V in volts, R in ohms, then P is in watts (W).

For a given voltage V, as with lamps in a lighting circuit, $P \propto 1/R$ from $P = V^2/R$. Hence a low filament resistance provides a larger power or light than a high filament resistance. If resistances are in series and not in parallel, the current I is the same; and thus, from $P = I^2R$, $P \propto R$ in this case.

EXAMPLE

A battery of e.m.f. 8 V and internal resistance 1 Ω is connected in series with coils of resistance 3, 4 and 5 Ω respectively. The 4 Ω coil is immersed in a liquid contained in a calorimeter and the rise in temperature of the liquid is found to be 1°C per minute. What would be the rate of rise in temperature if the 3 and 5 Ω resistances were connected in parallel in the circuit?

The current flowing, I, = $\dfrac{\text{e.m.f.}}{\text{total resistance}}$ = $\dfrac{8}{13}$ A.

The combined resistance, R, of 3 and 5 Ω in parallel is given by

$$\frac{1}{R} = \frac{1}{3} + \frac{1}{5}, \text{ from which } R = \frac{15}{8}\ \Omega.$$

$$\therefore \quad \text{total resistance} = 1 + 4 + \frac{15}{8} = 6\tfrac{7}{8}\ \Omega.$$

$$\therefore \quad \text{new current, } I_1, = \frac{8}{55/8} = \frac{64}{55}\ \text{A.}$$

For a given resistance, the heat per second $\propto$ current^2. Hence the new rate of rise in temperature, g, is given by

$$\frac{g}{1} = \frac{(64/55)^2}{(8/13)^2},$$

$$g = 3{\cdot}6°\text{C min}^{-1}.$$

High-tension transmission. From $P = IV$, a given amount of electrical power can be supplied at high current and low voltage, or at low current and high voltage. If the current is low, the power losses in cables of given resistance is low, since the power lost $= I^2R$. Thus high voltage (tension) transmission is adopted in the national grid system. Fig. 9.1b.

Hot-wire ammeter. Consists of a fine wire PQ joined to the terminals which sags slightly to M when current is passed. Fig. 9.2. The sag is taken up by a silk thread round a grooved wheel A, which then deflects the pointer. The scale is a square-law one, and as the heat produced $\propto I^2$, the instrument can be used to measure alternating current.

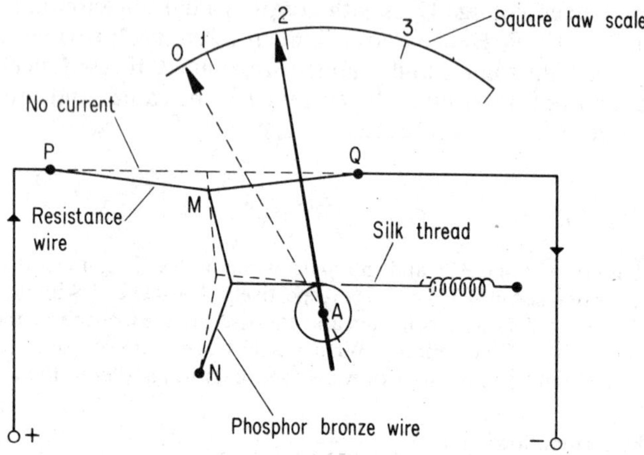

FIG. 9.2 Hot wire ammeter.

Maximum power from a battery. This is obtained when the external resistance = the internal resistance r (see p. 60). The maximum power supplied $= I^2R = E^2r/(r+r)^2 = E^2/4r$.

Units of energy and power. The *kilowatt-hour* (kWh) is the commercial unit. Since 1 kilowatt = 1000 W,

$$1 \text{ kWh} = 1000 \text{ W} \times 3600 \text{ s} = 3\cdot6 \times 10^6 \text{ J} = 3\cdot6 \text{ MJ (megajoules)}$$

Electromotive force and power. The e.m.f., E, of a cell may be defined as the *total power per unit current* obtainable from the cell. When the cell of internal resistance r is joined to a resistance R, and a current I flows.

$$\text{total power, } EI, = I^2R + I^2r.$$

$$\therefore \quad E = IR + Ir \text{ (p. 59).}$$

The power dissipated in the cell $= I^2r$; the power available $= I^2R$.

$$\therefore \quad \text{efficiency of cell} = \frac{I^2R}{EI} = \frac{R}{R+r}.$$

Thus the lower the internal resistance, the higher is the efficiency.

Measurement of resistance. If only an ammeter is available, the resistance R of a coil can be measured by immersing it in water in a calorimeter, and observing the temperature rise in a given time due to a current. Then, if H is the heat produced in joules, $H = I^2Rt$, or $R = H/I^2t$, with the usual notation. The current I in the circuit is kept constant, and its magnitude is observed on the ammeter.

If only a voltmeter is available, it is placed across the coil to measure its p.d. V in a circuit, and the heat H is determined as described above. Then, since $H = V^2t/R = V^2t/H$.

THERMO-ELECTRICITY

When two unlike metals are joined to form a complete circuit, and one junction is maintained at a different temperature from the other, a current flows. This is called the *Seebeck effect*. Fig. 9.3. The current flows from antimony to bismuth through the cold junction, and from iron to copper through the cold junction. A thermopile is a series arrangement of similar thermocouples, and is a sensitive detector of heat.

Measurement of thermo-electric e.m.f. A large resistance is placed in series with the potentiometer wire, to make the p.d. across the wire of the order of millivolts. This is shown in Fig. 8.6, p. 69.

Results. The e.m.f. E varies with the temperature $t°C$ of the "hot" junction according to

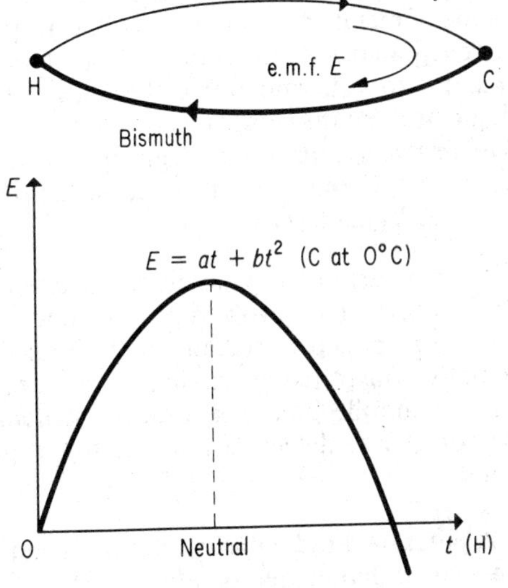

FIG. 9.3 Thermoelectric e.m.f. and temperature.

$$E = at + bt^2,$$

when the "cold" temperature is 0°C. Fig. 9.3. This is a parabolic variation. E has a maximum value at a temperature called the "neutral temperature", and reverses in direction when the temperature is increased sufficiently.

Measurement of temperature. A copper-constantan or iron-constantan thermocouple can be used from below 0°C to about 500°C. Up to 1200°C, a nickel-nichrome thermocouple can be used. Up to 1600°C, a platinum and alloy of platinium-rhodium thermocouple in a silica tube can be used. A millivoltmeter, calibrated in °C, is used with the thermocouple chosen. The thermoelectric pyrometer cannot be used for measuring temperatures beyond the neutral temperature, as identical e.m.f.s are obtained at two different temperatures in the complete parabolic variation of e.m.f.

The thermocouple is used to measure the melting-points of metals, or the temperature of a small area of a metal. It can be used to measure the *melting-point of naphthalene* by warning the latter until it all melts, immersing a thermocouple junction in the liquid, and then observing the reading in a sensitive galvanometer at regular intervals of time as the liquid cools. The flat portion of the curve corresponds to the melting-point, and is read from a previous calibration (deflection-temperature) curve with the same galvanometer.

Thermo-electric laws. (1) *Law of intermediate metals.* If two thermocouples are made of metals A, B and metals B, C respectively at the same hot and cold junction temperatures, $E_{AB} + E_{BC} = E_{AC}$, where E_{AC} is the e.m.f. due to a thermocouple of metals A, C at the same hot and cold junction temperatures.

(2) *Law of intermediate temperatures.* For a given thermocouple, $E_{12} + E_{23} = E_{13}$, where E_{12} is the e.m.f. at junction temperatures t_1, t_2°C, E_{23} at t_2, t_3°C, and E_{13} at t_1, t_3°C.

Peltier effect. When a current flows across the junction of two unlike metals, heat is absorbed or evolved at the junction. The effect reverses when the current is reversed, unlike the Joule heating. The Peltier effect can be demonstrated by passing a current of several milliamps through a thermopile and then quickly switching the thermopile to a galvanometer to detect current due to temperature change at the junctions.

Thomson effect. A potential gradient exists along a metal whose ends are maintained at different temperatures. The *Thomson coefficient* is the e.m.f. per unit temperature difference along the metal.

ELECTROLYSIS

Electrolysis. Definitions.   An *electrolyte* is a liquid which conducts an electric current. Salt and acid solutions are good electrolytes. Water is a very poor electrolyte. Electrolytes contain *ions*, which are atoms or groups of atoms carrying negative or positive charges of magnitude e, $2e$, or $3e$, where e is the charge on an electron. A hydrogen ion carries a charge $+e$; a hydroxyl ion (OH^-) carries a charge $-e$; a sulphate ion ($SO_4{}^{2-}$) carries a charge $-2e$. Generally, an ion carries a charge of magnitude ne, where n is the valency of the ion. An electron is denoted by the symbol e^-.

The *cathode* is the plate where the conventional current leaves the electrolyte; the *anode* is the plate where the conventional current enters the electrolyte. The "metal" is always deposited on the cathode.

Electrolysis of acidulated water. With platinum electrodes A, C, hydrogen ions drift to the cathode C and are deposited. Negative sulphate and hydroxyl ions drift to the anode A, and the hydroxyl ion is discharged at A in preference to the sulphate ion as its discharge potential is smaller. Oxygen is produced: $4\,OH^- = 2H_2O + O_2 + 4e^-$.

Electrolysis of copper sulphate solution. With copper electrodes A, C, hydrogen and copper ions drift to the cathode C; the copper ions are discharged in preference to the hydrogen ions as the discharge potential of the former is smaller.

The negative hydroxyl and sulphate ions drift towards the anode A. The copper ions from the anode go into solution, however, as the discharge potential of the copper ions is less than that of the hydroxyl and sulphate ions. Thus copper is lost from the anode and an equal amount is deposited on the cathode.

Faraday's laws of electrolysis. *First law.* The mass m of an element liberated in electrolysis is proportional to the current I and the time t. Fig. 9.4a. Explanation: The charge $Q(= I.t)$ is carried through the electrolyte by their ions; hence mass deposited $\propto Q \propto I.t$. Fig. 9.4b.

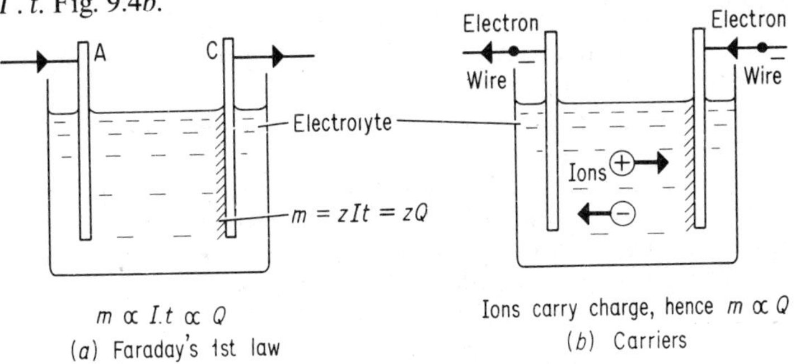

$m \propto I.t \propto Q$

(a) Faraday's 1st law

Ions carry charge, hence $m \propto Q$

(b) Carriers

FIG. 9.4 Ion drift and Faraday's first law.

Second law. For a given current and time, the mass of an element deposited in a voltmeter is proportional to its *relative atomic mass/valency* (chemical equivalent). Explanation: An ion carries a charge $+e$, $+2e$ or $+3e$, if its valency is 1, 2 or 3 respectively. Thus a given charge of $6e$ is carried by 6 monovalent ions or 3 divalent ions or 2 trivalent ions. The ratios of the masses deposited would be respectively $6A_1 : 3A_2 : 2A_3$, where A_1, A_2, A_3 are the respective relative atomic masses. This ratio $= A_1/1 : A_2/2 : A_3/3$, or "relative atomic mass/valency" for each element. Fig. 9.5.

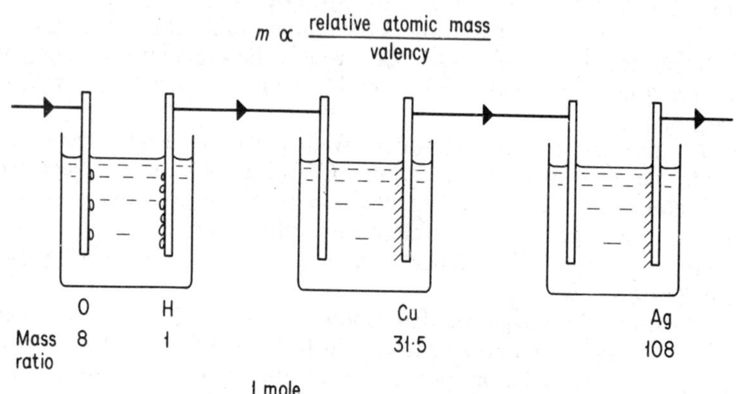

FIG. 9.5 Faraday's second law.

Electrochemical equivalent. The electrochemical equivalent (e.c.e.) of an element is the mass deposited by 1 ampere in 1 second.

From the first law,

$$m = zIt,$$

where z is the e.c.e. of the element. Since $It = Q$, the quantity of electricity, z is expressed in "kilogrammes per coulomb", kg C^{-1}.

The e.c.e. of hydrogen, copper and silver are respectively $1\cdot04 \times 10^{-8}$, $3\cdot3 \times 10^{-7}$, $1\cdot118 \times 10^{-6}$ kg C^{-1}.

From Law 2, *the e.c.e. is proportional to the relative atomic mass/valency (chemical equivalent).*

The *Faraday constant* (F) is the charge carried by a mole of any monovalent element, and is 96500 coulombs (approx.). The number of positive or negative ions in the mole = the number of molecules = $6\cdot02 \times 10^{23}$ (N_A), = Avogadro constant. Thus each ion carries a charge $= 96500C/6\cdot02 \times 10^{23} = 1\cdot6 \times 10^{-19}C$ (approx.). This is the charge on an electron, i.e. $N_A e = F$.

The current I is given by $I = m/zt$. Thus if z is accurately known, as in the case of silver, the current in a circuit can be found accurately by including a silver voltameter. In laboratory practice a copper voltameter may be used.

EXAMPLE

A copper and a silver voltameter are in series in a circuit. If 0·80 g of copper is deposited in 20 min, calculate (*i*) the current flowing, (*ii*) the mass of silver deposited in the same time. (Assume relative atomic masses of copper and silver = 64 and 108, 1 Faraday = 96 500 C.)

(*i*) $$\text{For copper,} \quad \frac{\text{relative atomic mass}}{\text{valency}} = \frac{64}{2} = 32$$

$$\therefore \quad 32 \text{ g is deposited by } 96\,500 \text{ C}$$

$$\therefore \quad 0\cdot8 \text{ g is deposited by } \frac{0\cdot8}{32} \times 96\,500 = 2412\cdot5 \text{ C}$$

$$\therefore \quad I = \frac{Q}{t} = \frac{2412\cdot5}{20 \times 60} = 2\cdot0 \text{ A.}$$

(*ii*)   For silver, relative atomic mass/valency = 108/1 = 108.
   Hence, from Faraday's second law, mass m of silver is given by

$$\frac{m}{0\cdot8} = \frac{108}{32}, \quad \text{or} \quad m = 2\cdot7 \text{ g.}$$

Determination of e.c.e.   (1) *Copper.* Copper voltameter, with copper electrodes in acidulated copper sulphate solution, used. A "dummy" cathode is first used to obtain a suitable current, for example 0·5 to 1·0 A. A clean cathode is required, and it is carefully washed and dried after the experiment.

(2) *Hydrogen or oxygen.* Water voltameter of Hoffmann type used. The pressure of the gas is the observed pressure less the saturation vapour pressure (s.v.p.) of water at the temperature concerned. By reducing the volume of the gas to s.t.p., and using the density value at s.t.p., the mass of the gas is calculated.

CELLS

TYPE	POSITIVE POLE	NEGATIVE POLE	ELECTRO-LYTE	DEPOLARI-ZER	SPECIAL FEATURES
Daniell (E.m.f. — 1·08 V, *r* about 1 ohm.)	Copper.	Zinc.	Dil. H_2SO_4.	$CuSO_4$ soln.	Maintains a small current for a long time.
Leclanché (E.m.f. — 1·45 V, *r* about a few ohms.)	Carbon.	Zinc.	Dil. NH_4Cl.	MnO_2.	Polarizes rapidly; used in bell circuits.
Weston cadmium (Fig. 9.6). (E.m.f. — 1·0186 V at 20°C; *r* about a few hundred ohms.)	Mercury.	Cadmium amalgam.	$CdSO_4$ solution.	Hg_2SO_4.	Standard cell. Used in potentiometer experiments to calibrate wire.

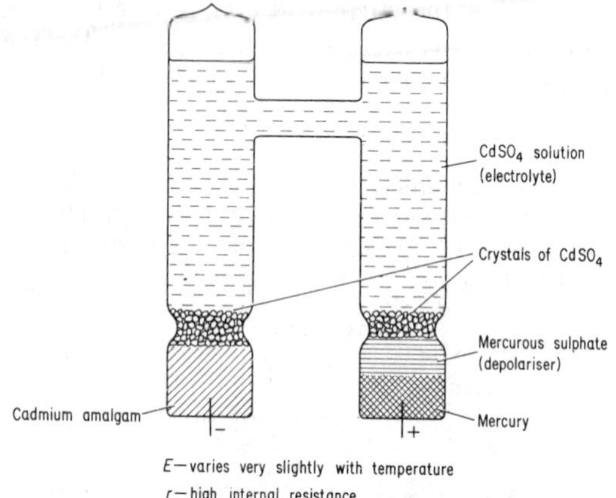

FIG. 9.6 Standard cell—Weston cadmium.

Lead-acid accumulator.   Maintains a large current for a long time. The positive pole is lead peroxide (PbO_2); the negative pole is lead (Pb); the electrolyte is sulphuric acid of relative density 1·25 when the cell is freshly made. The e.m.f. is then about 2 volts, and the internal resistance about 0·01 ohm.

Discharging.   When used, a little lead sulphate forms at both plates, the relative density of the acid diminishes (1·18V minimum) and the e.m.f. slowly decreases (1·9V minimum). At the positive plate: $PbO_2 + 2H^+ + H_2SO_4 + 2e^- = PbSO_4 + 2H_2O$. At the negative plate: $Pb + SO_4^{2-} = PbSO_4 + 2e^-$.

Charging.   The current is passed into the accumulator from the positive to the negative plate, the d.c. mains being used. A series resistor is essential to limit the current to the magnitude required, as the resistance of the acid is very low. Charging is stopped when the relative density rises to 1·25 again, and "gassing" takes place. The e.m.f. may then have risen to 2·7V, but it settles at about 2V soon after use.

Nickel-alkaline (Nife) accumulator.   This is more robust than the lead-acid type, and is not easily damaged by excessive discharge or charge, but has an e.m.f. of about 1·25 volts. The positive plate is nickel hydroxide, the negative plate is a mixture of iron and cadmium, and the electrolyte is caustic potash solution of relative density about 1·17.

EXAMPLE

A battery of 30 accumulators is to be charged at 4 A from 240 V d.c. mains. How may this be done? What percentage of power will be wasted? (Assume that the mean voltage of each accumulator during charging is 2·2 V and that its internal resistance is 0·1 Ω.)

The accumulators are arranged in series, and the positive end is connected to the + terminal of the mains, the negative end to the − terminal. A suitable variable resistance R is placed in series in the circuit whose value is calculated as follows:—

$$I = \frac{\text{net e.m.f.}}{\text{total resistance}} = \frac{240 - 30 \times 2 \cdot 2}{R + 30 \times 0 \cdot 1}$$

$$\therefore \quad 4 = \frac{174}{R + 3}$$

$$\therefore \quad R = 40 \cdot 5 \ \Omega$$

The total power supplied $= IE = 4 \times 240 = 960 \ \text{W}$

The power wasted $= I^2 R + I^2 r = I^2 (R + r)$

$$= 4^2 (40 \cdot 5 + 3) = 696 \ \text{W}$$

$$\therefore \quad \text{percentage of power wasted} = \frac{696}{960} \times 100 = 72 \cdot 5.$$

10. ELECTROMAGNETISM. FORCE ON CONDUCTOR

Magnetic flux density. Symbol: B. Unit: tesla, T.

Definition of B: Obtained from the force on a conductor in the magnetic field, i.e. $B = F/Il$. Direction of B: This is the direction of a current carrying conductor when no force acts on it in the field. See Fig. 10.1a. Otherwise, the direction of B may be found by

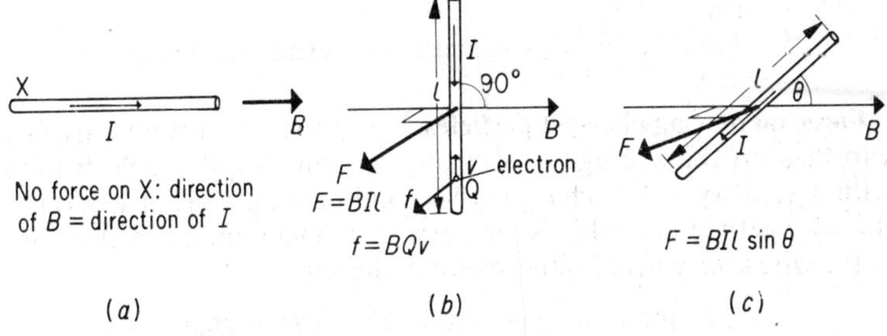

FIG. 10.1 Force on conductors

applying Fleming's left hand rule (B — fore-finger, I — middle finger, F — motion — thumb.

Note The unit of flux, Φ, is the weber, symbol Wb. $1T = 1 \ \text{Wb m}^{-2}$.

General formulae. $\qquad F = BIl \sin \theta,$ $\qquad$ . . . . (1)

where θ is the angle between the conductor and the field B. Fig. 10.1c.

$$\therefore \quad F = BIl, \qquad \text{. . . . . (2)}$$

when the conductor is perpendicular to the field B. Fig. 10.1b. Units: F in newton (N) when B in tesla (T), I in ampere (A), l in metre (m).

Experiment. To demonstrate the factors in (2) on which F depends, place a wire X inside a long solenoid perpendicular to its axis and counterbalance its weight. Vary B by altering the current in the solenoid. Vary I by altering the current in the wire.

Field interaction. Fig. 10.2a, b, c shows the magnetic field B, current field of A, and combined or resultant field. Note A moves from a region of high to low flux-density in the resultant field.

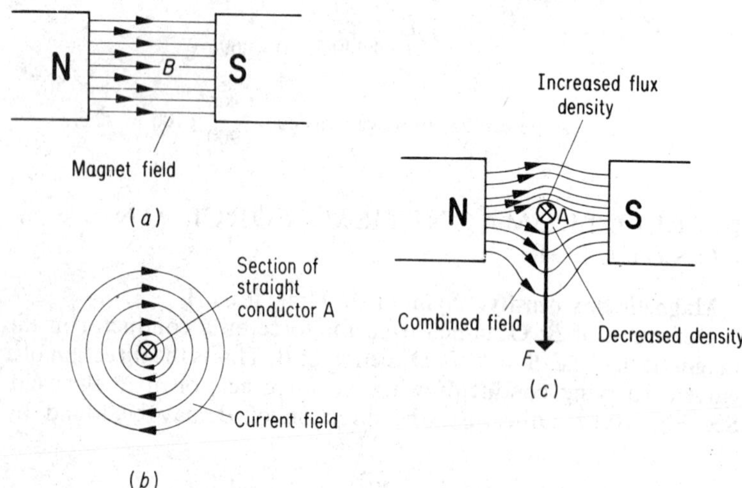

FIG. 10.2 Field interactions and Force on conductor.

Force on moving charged particles. Suppose a current-carrying wire has on the average n electrons per unit length, each drifting with a velocity v. The charge per second passing a given section of the wire is then nev. This is the current, I. The number of electrons in the wire is nl, where l is the length of the wire.

$$\therefore \quad \text{force on } nl \text{ electrons } F = BIl = Bnevl$$

$$\therefore \quad \text{Force per electron, } F = \frac{Bnevl}{nl} = Bev \qquad \text{. . (1)}$$

Hall effect. A perpendicular magnetic field B, applied to a current carrying conductor, produces a *transverse e.m.f.*

Explanation. With electron carriers, the force on each $= Bev$, in a direction transverse to the conductor. Fig. 10.3. The electron drift ceases when $Bev = Ee$, where $E =$ electric intensity $= V_H/d$, V_H being the Hall voltage and d the width of the conductor. Hence $V_H = Bvd$. Since $I = nevA$ (p. 54) $= nevdt$, where t is the thickness, then $V_H = BI/net$.

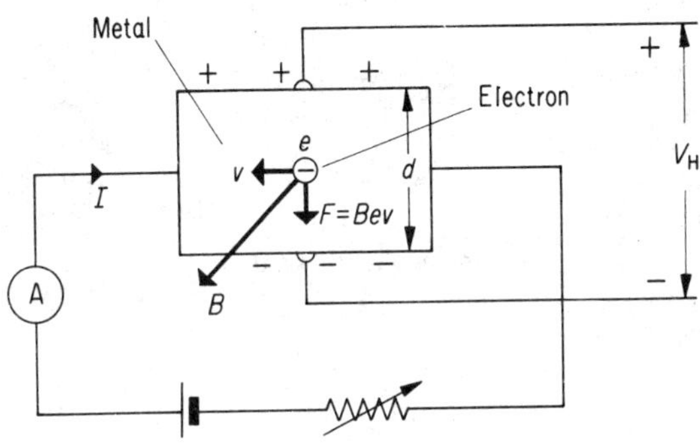

Fig. 10.3 Hall effect in metal.

Applications. (1) Hall probe for measuring B ($B \propto V_H$) (2) Detecting sign of current carriers—sign of Hall voltage depends on sign of charge on current carriers, and experiment shows that *electrons* (negative charges) are carriers in metals and *holes* are majority carriers in p-semi conductors (p. 133).

Torque (Couple) on Current-carrying Coil. (1) With the plane parallel to the field B,

$$\text{torque or couple, } C = NABl$$

(2) With the *normal* to the plane of the coil at an angle θ to B, Fig. 10.4*a*, the forces F produce torque C but the forces F_1 cancel.

$$C = NABl \sin \theta$$

Note that a current-carrying coil acts as a magnet whose axis lies along the normal to the plane of the coil. Thus

$$\text{magnetic moment, } m, \text{ of coil} = NAI. \quad \text{Unit of } m : \text{A m}^2.$$

The torque on a coil of any shape is given by $NABl \sin \theta$. Fig. 10.4*b*. In the special case of a coil in a *radial field*, as in the moving-coil meter, the torque $= NABl$, since the plane of the coil is always parallel to the field in which it is situated Fig. 10.4*c*.

81

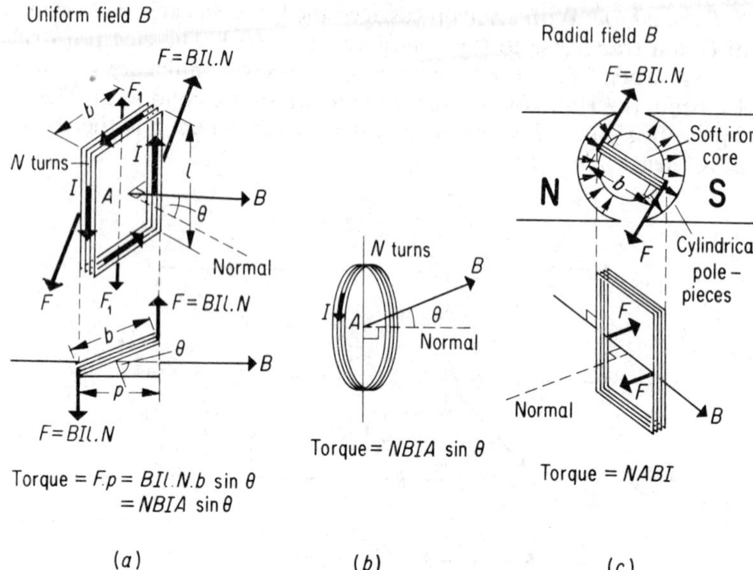

FIG. 10.4 Torque (Couple) on coil in uniform and radial fields.

EXAMPLE

A rectangular coil of 50 turns hangs vertically in a uniform magnetic field of 10^{-2} T, so that the plane of the coil is parallel to the field. The mean height of the coil is 5 cm and its mean width 2 cm. Calculate the strength of the current that must pass through the coil in order to deflect it 30° if the torsional constant of the suspension is 10^{-9} N m per degree.

When the coil is deflected 30°, the vertical sides of the coil, 5×10^{-2} m long, are perpendicular to the field.

$$\therefore \quad \text{force on each side, } F, = NBIl = 50 \times 10^{-2} \times I \times 5 \times 10^{-2} \text{ newton}$$

The perpendicular distance between the forces $= b \sin 60° = 2 \times 10^{-2} \sin 60°$ metre

$$\therefore \quad \text{moment (torque) of deflecting couple}$$

$$= 25 \times 10^{-3}I \times 2 \times 10^{-2} \sin 60° \text{ N m}$$

$$= 5 \sin 60° \times 10^{-4}I \text{ N m}$$

$$\therefore \quad 5 \sin 60° \times 10^{-4}I = 30 \times 10^{-9}$$

$$\therefore \quad I = 7 \times 10^{-5} \text{ A.}$$

Magnet. Torque or couple on magnet inclined at θ to field $B = mB \sin \theta$ (p. 81).

$$\therefore \quad \text{Period of oscillation, } T = 2\pi \sqrt{\frac{K}{mB}},$$

where K = moment of inertia. Thus for an *oscillating magnetic needle*,

$$B \propto \frac{1}{T^2} \propto f^2,$$

where f is the frequency of vibration.

Comparison of fields. $B_1/B_2 = f_1^2/f_2^2$, using an oscillating magnetic needle (see p. 101).

Moving-coil instrument. Consists of (*i*) rectangular coil, (*ii*) soft-iron cylinder between curved pole-pieces of a permanent magnet, (*iii*) springs to control the rotation, (*iv*) a linear scale. The coil is wound on an aluminium former to make it come to rest quickly when it is deflected—this is due to eddy currents circulating in the aluminium, p. 91. The coil is pivoted in jewelled bearings to reduce friction, and the current is led in and out of the coil to the terminals via the springs. (Fig. 10.5*a*).

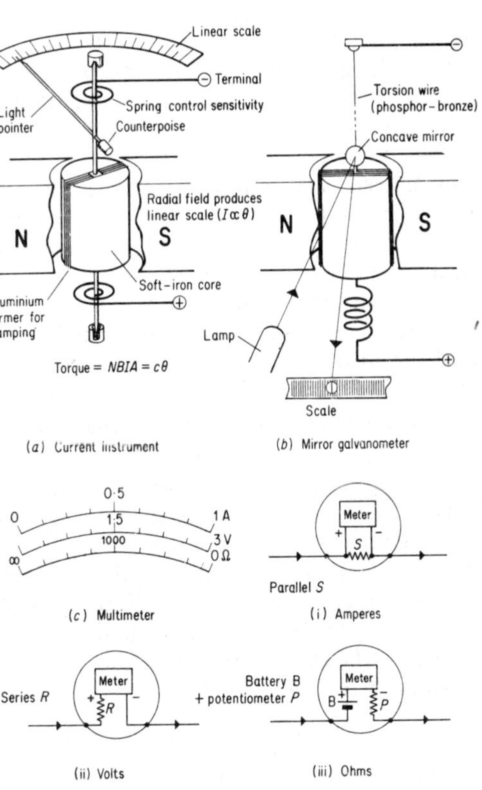

FIG. 10.5 Moving-coil instrument. Multimeter.

Theory. The soft-iron cylinder provides a radial magnetic field. When the coil is deflected, therefore, its plane is *parallel* to the magnetic field B in which it rests. The forces, BIl, on each of the two vertical sides are perpendicular to the field B, and hence are perpendicular to the plane of the coil. Thus the distance between the forces $= b$, the breadth of the coil. This is *always* the case when the coil comes to rest in a radial field. (In a non-radial field such as a uniform field, the distance between the forces $= b \sin \theta$, p. 82, and hence varies with θ).

For equilibrium of the coil, if c is the opposing couple per radian twist of the spring, deflecting couple on coil = opposing couple due to torsion in spring.

$$\therefore \quad NABI = c\theta.$$

$$\therefore \quad I = \frac{c}{NAB} \cdot \theta, \qquad \text{or} \quad I \propto \theta$$

Thus the current is proportional to the angle of deflection of the coil, that is, the scale is a linear or uniform one.

The *sensitivity* of the instrument is the *deflection per unit current*, or θ/I. Hence, from the relation between I and θ, sensitivity $= NAB/c$. A low value of c will thus produce high sensitivity. This is the case for the mirror galvanometer.

Mirror galvanometer. The ammeter and milliammeter have springs to control the coil's rotation. In the mirror galvanometer the springs are replaced by a fine phosphor-bronze torsion wire, which has a weaker control (Fig. 10.5b). The instrument can then be used to measure tiny currents of the order of fractions of microamps. Instead of a pointer, a light beam is used. This is incident on a mirror of large radius of curvature, usually 1 metre, attached to the coil, and reflected to a millimetre translucent scale 1 metre away.

Conversion of milliammeter to ammeter, voltmeter, ohm-meter (multimeter). Fig. 10.5c.

Ammeter. Add a shunt, i.e. a resistance S in *parallel* with the milliammeter. The shunt diverts most of the current through itself, and has thus a *low resistance* compared with the milliammeter resistance G. If an ammeter of 0—1 A is required, and the milliammeter reads 0—5 mA, then 5 mA or 0·005 A flows through the meter (full scale deflection) when 0·995 A flows through the shunt. Since the p.d. is the same across the meter terminals either through the meter or through the shunt, from Ohm's law,

$$0·995 \times R = 0·005 \times G, \qquad \text{or} \quad R = G/199.$$

Voltmeter. Add a *series* resistance R of high resistance. If a voltmeter 0—3V is required using the 0—5 mA instrument, then a p.d. of 3V produces a current in the meter of 5 mA or 0·005 A.

$$\therefore \quad \text{total resistance needed} = \frac{3 \text{ V}}{0 \cdot 005 \text{ A}} = 600 \text{ ohms.}$$

Hence series resistor $R = 600 - G$.

Ohm-meter. Connect a small dry battery B of a few volts in series with the coil and a high resistance rheostat ("pot") P. With the meter leads touching, adjust P until a full-scale deflection is obtained. This is the "zero" of the *resistance* scale. Now place varying known resistances between the leads and note the current each time. In this way the scale can be calibrated in "ohms". An unknown resistance can then be read directly.

Resistance and sensitivity of mirror galvanometer. By using a potential divider, a small p.d. V can be applied to the galvanometer G in series with a resistance box of known values R. Then $I = V/(R+G) = c\theta$.

$$\therefore \quad R + G = \frac{V}{c}\left(\frac{1}{\theta}\right)$$

A straight line graph of R v. $1/\theta$ thus cuts the $1/\theta$-axis at an intercept value of $-G$, and hence the galvanometer resistance can be found.

From a knowledge of G, the current value I $[= V/(R+G)]$ can be calculated in microamperes. The corresponding scale deflection, x millimetres, is known.

$$\therefore \quad \text{sensitivity} = x/I \quad \text{mm } \mu A^{-1}$$

Ballistic galvanometer. This is a moving-coil instrument, but (*i*) the moving-system is heavy, so that it moves only very slightly when a discharge takes place through the instrument, (*ii*) there is no damping as no former is present. (The former is removed after the wires are covered with shellac.) The coil is then subjected to an impulse when a quantity Q of electricity passes quickly through it, and calculation shows that $Q \propto \theta$, the *first throw* of the needle. The ballistic galvanometer is used to measure capacitances and flux change (pp. 48 and 97).

Force between currents. If two parallel wires carry currents in the *same* direction, they are *attracted*. In opposite directions, repulsion occurs. The direction of the force F in either case can be found by applying Fleming's left-hand rule to one of the wires. Fig. 10.6 shows the field between the parallel wires.

The force per metre on a wire carrying a current I in a perpendicular field $B = BI$ (p. 80). The flux-density B due to the second wire carrying a current $I_1 = \mu_0 I_1/2\pi r$, where r is the distance apart of the wires (p. 104). Thus

$$\text{force per metre on either wire} = \mu_0 I I_1/2\pi r \quad . \quad . \quad . \quad (1)$$

Units: Force per metre in N m^{-1} when I, I_1 in A, r in m, μ_0 = $4\pi \times 10^{-7}$ H m^{-1}

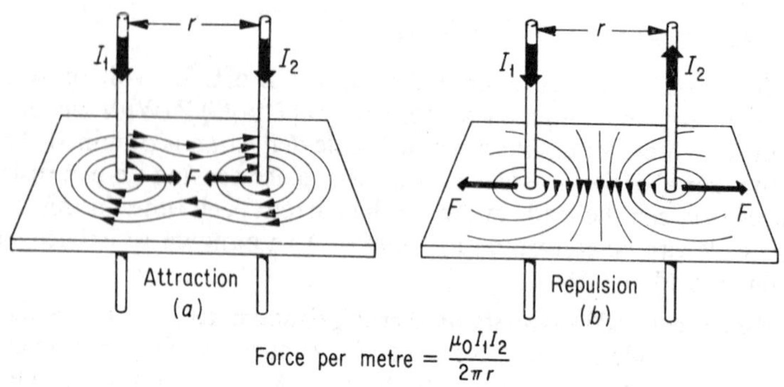

$$\text{Force per metre} = \frac{\mu_0 I_1 I_2}{2\pi r}$$

FIG. 10.6 Force between currents.

Definition of ampere.   The *ampere* is defined as that current which, flowing in two straight parallel conductors of infinite length and negligible circular cross-section and placed 1 metre away from each other in a vacuum, produces between the conductors a force of 2×10^{-7} newton per metre length.

Substituting $I = I_1 = 1$ and $r = 1$ in the expression (1), and using force per metre $= 2 \times 10^{-7}$ N, we find $\mu_0 = 4\pi \times 10^{-7}$ H m^{-1}.

Ampere balance.   *Laboratory form.* Fig. 10.7a. CL is repelled by HG, so $F = mg \propto I^2$.

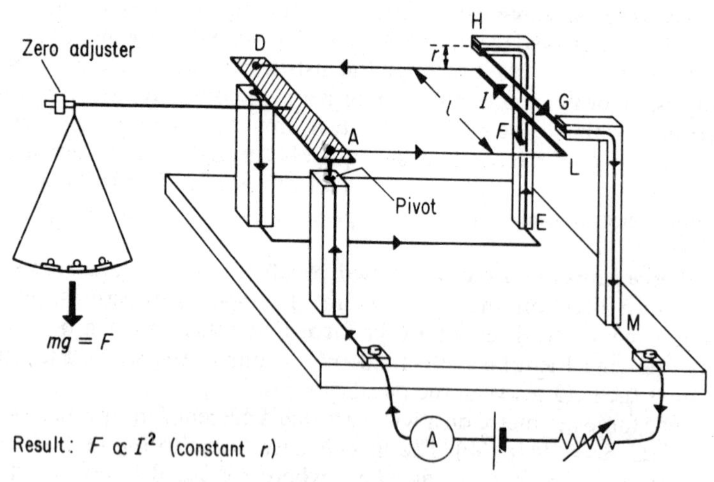

(a) Laboratory investigation

FIG. 10.7. Ampere balance.

86

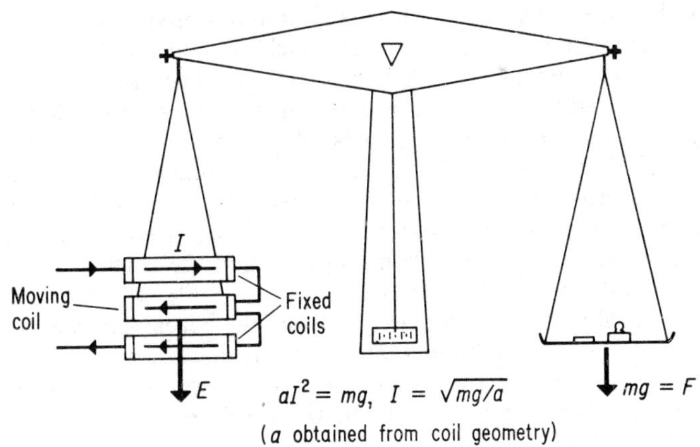

FIG. 10.7 Ampere balance.

Accurate form. Fig. 10.7b. Middle coil moves downward. $F = mg = aI^2$.

Wattmeter. Basically, this has (i) a "voltage coil" M which can rotate, (ii) "current coils" F, F which are fixed and provide a magnetic field B for the voltage coil. Fig. 10.8. The torque or couple which deflects the voltage coil is proportional to the product BI', since torque $= NABI'$.

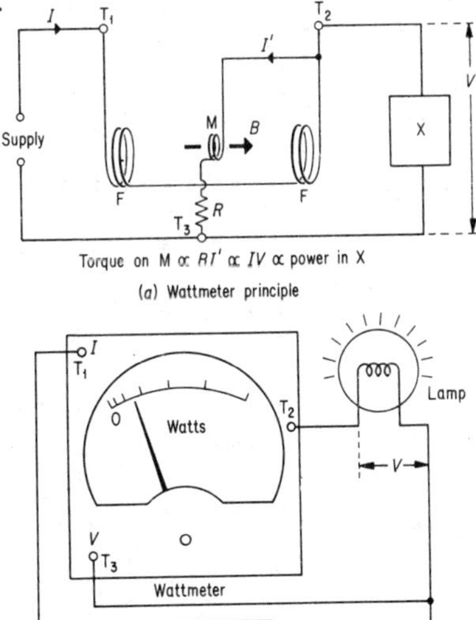

FIG. 10.8 Wattmeter principle.

Now $B \propto I$, the current in the machine monitored, and $I' \propto V$, the p d across the terminals of the machine. Thus $BI' \propto IV \propto$ power. Springs control the rotation of the voltage coil, and the scale is calibrated from a previous experiment in watts or kilowatts.

Simple D.C. motor. This consists of a rectangular coil in a magnetic field. The terminals of the coil are connected to a split-ring commutator which revolves with the coil, and which presses against brushes connected to the d.c. supply. From Fleming's left-hand rule the coil rotates continuously, and the current in the coil reverses after every half-revolution.

Series-wound motor. In practical motors the permanent magnet is replaced by an electromagnet with a coil round it known as the *field winding*, or field; the rotating coil is wound round a soft-iron core in many planes and is known as the *armature*; and the commutator consists of many insulated segments.

In the series-wound motor the same current flows through the armature and the field. The speed drops considerably as the load attached to the motor increases. The initial pull of the series-wound motor is high, and it is therefore used in cranes and electric trains, where a large initial starting torque is required.

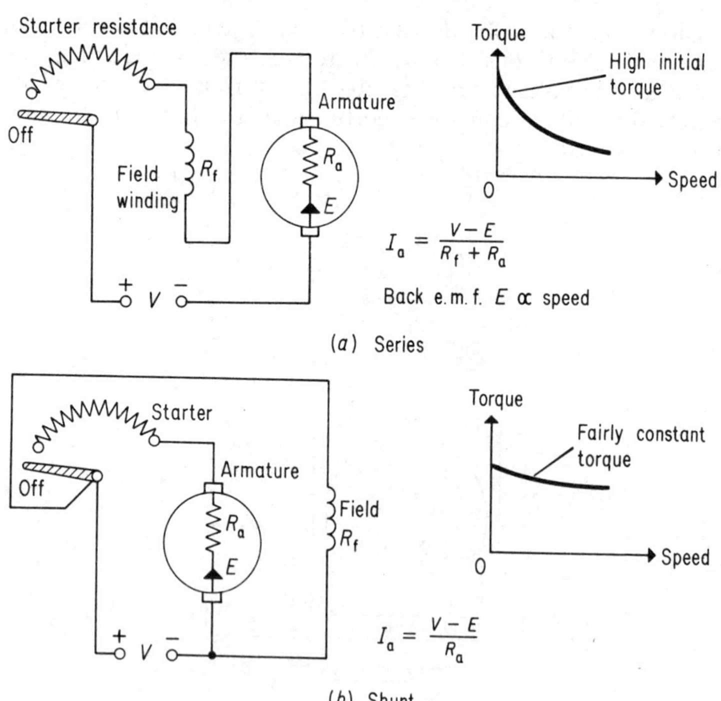

(a) Series

(b) Shunt

Fig. 10.9 Electric motors.

Back e.m.f. When the motor is running, the armature cuts magnetic flux continuously. An induced e.m.f. *e*, proportional to the speed, is set up which opposes the applied e.m.f. *V*, and thus the armature current I_a is given by $(V-E)/(R_a+R_f)$, where R_a, R_f are the resistances of armature and field windings. Fig. 10.9a. When the current is first switched on $E = 0$. A very high current would flow in this case, ruining the motor. Hence a *starter resistance* is used initially, and slowly cut out as the motor gains speed. If the speed drops when a load is connected, the armature current increases because the back e.m.f. then decreases. This provides the increased power necessary to drive the load.

Shunt-wound motor. In this case the field is in parallel with the armature. The field current is therefore constant. The armature current I_a is given by $I_a = (V-E)/R_a$, or $E = V-I_aR_a$. Fig. 10.9b. The back e.m.f. is proportional to the speed.

The speed of this motor drops only slightly as the load is increased. The motor is therefore used for driving machines.

11. ELECTROMAGNETIC INDUCTION

Flux change. The flux linking a *coil* when the flux enters all the *N* turns normally is given by

$$\Phi = NAB$$

When *B* is in T and *A* in m², then Φ is in Wb. If the coil is turned so that the normal to the plane makes an angle θ with *B*, then, Fig. 11.1a,

$$\text{flux linkage } \Phi = NAB \cos \theta.$$

When a *straight conductor* moves through a field and cuts the flux normally. Fig. 11.1b, the flux "cut" $\Phi = B.A. = B.l.h$, where *l* is the length of conductor and *h* is the distance moved.

Lenz's law. The direction of the induced current opposes the motion or change producing it.

Faraday's law. The induced e.m.f. is proportional to the rate of change of the magnetic flux linking the circuit.

Deduction and verification of Lenz's law. Lenz's law follows from the Principle of the Conservation of Energy. Thus suppose the current flows clockwise when the N pole of a magnet approaches a coil. The attractive force on the magnet makes it move faster, which increases the induced current strength, which now increases the speed of the magnet further. Thus the kinetic energy of the magnet, and the electrical energy in the coil, both increase without the expenditure of work. This is contrary to the Principle of the Conservation of Energy. Thus the current flows *anti-clockwise*, not clockwise as assumed. In this case work is done against the opposing force when the magnet is moved, and the energy reappears as electrical energy in the coil.

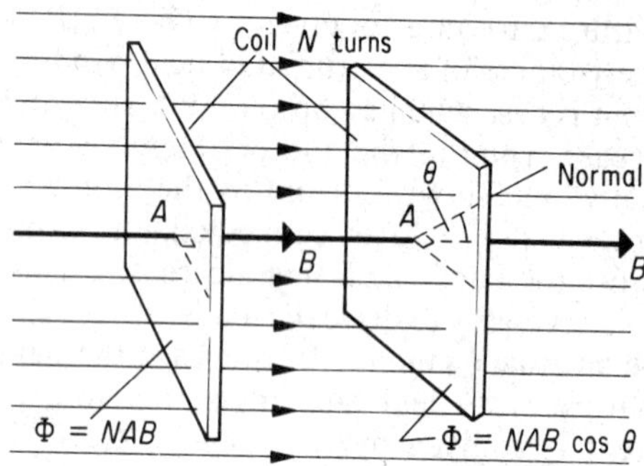

(a) Flux linkage, Φ

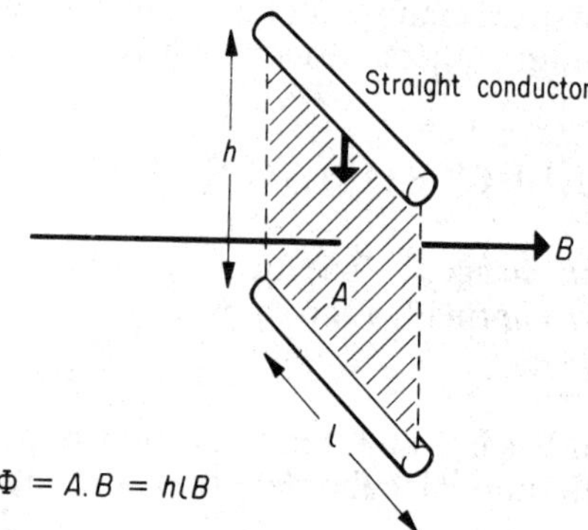

(b) Flux cutting

FIG. 11.1 Flux linkage and cutting.

Lenz's law can be verified by connecting a centre-zero galvanometer to a solenoid, and plunging the north pole of a magnet in the coil. The needle may then move to the right. The circuit is now broken, a large resistance and accumulator is added, and the needle deflection is arranged to be to the right. A compass-needle will then show that the face of the coil acts like a north pole. Thus the induced current opposed the movement of the north pole of the magnet in the first case.

Verification of Faraday's law. An earth inductor coil is connected to a galvanometer. The coil is rotated in the earth's field at five different speeds, and the corresponding steady deflection θ is obtained each time. The number of revolutions per minute is plotted against θ. A straight line graph passing through the origin is obtained, showing that the induced current, and hence e.m.f., is proportional to the rate of change of the flux.

Induction coil. Consists of (*i*) *primary coil* with a few hundred turns connected to a low-voltage d.c. supply with a *make-and-break device*, (*ii*) a *secondary coil* of several thousand turns, (*iii*) a large *capacitor* across the make-and-break, (*iv*) a *soft-iron core* in the form of iron strips. The make-and-break produces a rapid change of flux in the secondary at the break, and thus a high induced voltage of some thousand volts at this instant. The induced voltage at the make is low as the flux change is much slower. Thus, on the average, a high d.c. voltage is produced at the secondary terminals (Fig. 11.2*a*).

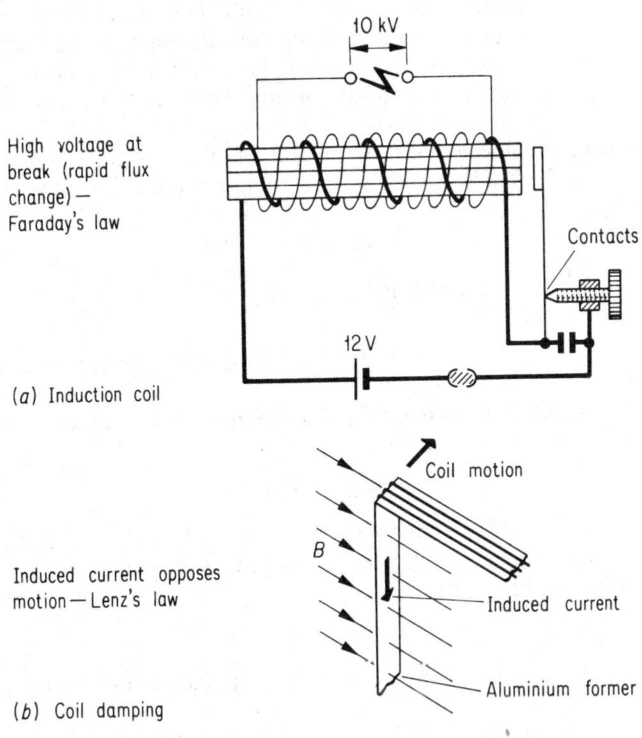

Fig. 11.2 Induction coil. Eddy currents.

Induction experiments. *Arago's disc.* When a disc of copper is rotated above a pivoted magnet, the latter begins to rotate in the same direction. Induced (eddy) currents are set up in the copper which opposes the motion, and thus the magnet rotates in the same direction. Conversely, a magnet spinning very fast in front of a copper disc makes the disc rotate in the same direction; an aluminium disc of the same size does not rotate as fast, since its resistance is higher and the induced current is smaller.

Moving-coil galvanometer. The coil is wound on a metal (aluminium) frame, Fig. 11.2b. As the coil oscillates in the deflected position, it cuts the flux in the radial field, and an induced or eddy current is obtained. This opposes the motion, from Lenz's law, and the coil therefore comes quickly to rest. In a ballistic galvanometer, which measures quantity of electricity, the coil is wound on a light wooden frame, or no frame, so that no eddy current flows. The galvanometer is therefore undamped in this case.

Eddy-current losses. Eddy currents flow in the iron cores of an induction apparatus, and cause heat. The loss of energy is considerably reduced by making the core into laminated sheets, with varnish between to insulate one sheet from the next, as the induced currents now no longer circulate in the metal. A bundle of iron wires may be used as the induction coil core, as the wires touch only at a relatively few points.

Formula for induced e.m.f. If a coil of N turns and area A is situated in a field of flux-density B so that all the flux links the turns normally, then

$$\text{flux-linkage } \Phi = NAB$$

$$\therefore \quad \text{induced e.m.f., E} = -\frac{d\Phi}{dt} = -NA\frac{dB}{dt}$$

Units: E in volts (V) when B in tesla (T), A in metre^2 (m^2) and t in second(s).

Induced e.m.f. in moving straight conductor. The induced e.m.f., E, is given by

$$E = Blv \sin \theta,$$

where l is the length, v is the velocity, B is the flux-density and θ is angle between the conductor and B. When the conductor moves at 90° to the field, Fig. 11.3a

$$E = Blv.$$

Units: E in volts when B in tesla (T), l in metre (m), v in metre second^{-1} (m s^{-1}).

Explanation of formula. When a straight conductor moves with a velocity v perpendicular to a field B, a *force* acts on the moving charges or electrons equal to Bev per electron (p. 80). This force pushes the electrons to one end Y of the conductor, which then has a negative potential relative to the other end X. Fig. 11.3a. The motion ceases when $E'e = Bev$, where E' is the electric intensity due to the charges at X and Y. Hence $E' = Bv$. But $E' = $ potential gradient $= E/l$. Hence $E = Blv$.

Direction of e.m.f. Fleming's RIGHT-hand rule states: If the first three fingers of the right hand are held mutually at right angles, the forefinger in the field direction, the thumb in the direction of motion, then the middle finger points in the direction of the induced current (*I*).

Note particularly that the conductor is a *generator* of e.m.f., like a battery. Thus the middle finger in Fleming's rule points towards the end of the conductor at a *higher* potential than the other. This is the end X of the conductor XY mentioned previously.

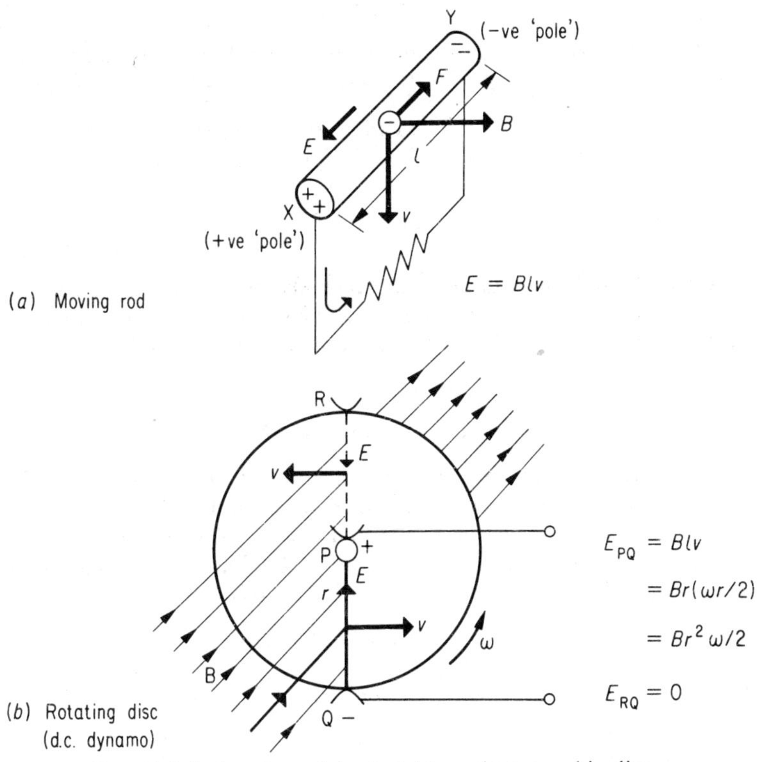

Fig. 11.3 Induced e.m.f. in straight conductor and in disc.

Disc dynamo. When a metal disc is placed between the poles of a magnet so that the magnetic flux is perpendicular to the plane of the disc, and then rotated steadily, an induced e.m.f. *E* is obtained between the centre of the disc and one edge. Fig. 11.3b. The radius between the centre and edge of the disc is a straight conductor moving with an average velocity $v = r\omega/2$, where $\omega = 2\pi f$ and *f* is the number of revs per second. Thus

$$E = Blv = B.r.r.2\pi f/2 = B.f.\pi r^2$$

Note that there is no e.m.f. between opposite ends of a *diameter*; the two radii from the centre are moving in opposite directions at any instant and hence their e.m.f.s cancel.

EXAMPLE

A copper disc of radius 15·0 cm rotates 25 times per second with its plane perpendicular to a uniform magnetic field. If an e.m.f. of 5 millivolts is induced between the centre and edge of the disc, determine the flux density of the magnetic field.

The induced e.m.f. E across the radius is given by

$$E = Blv,$$

where v is the *average* velocity of the radius. From velocity $= r\omega$, the average velocity $= r\omega/2$, where $r = 15 \times 10^{-2}$ and $\omega = 2\pi f = 2\pi \times 25$. Since $l = r$,

$$\therefore \quad \frac{5}{1000} = \frac{B \times 15^2 \times 10^{-4} \times 2\pi \times 25}{2}$$

$$\therefore \quad B = 28 \times 10^{-4} \text{ T.}$$

Induced e.m.f. in rotating coil. Simple dynamo. When a coil of N turns and area A rotates at a constant angular speed ω in a uniform field B, the induced e.m.f. at any instant is given by

$$E = E_0 \sin \omega t, \quad \ldots \ldots \ldots \ldots \quad (i)$$

where $\quad E_0 = \text{maximum e.m.f.} = \omega NAB$

Units: E_0 in volt when $\omega(= 2\pi f)$ in rad s^{-1}, A in metre^2, B in T.

Proof: Consider the rotating coil at an instant and let $\theta = $ angle made with B by the *normal* to the plane of the coil, i.e. $\theta = 0°$ when the plane is perpendicular to B. Fig. 11.4a. Then flux linking coil normally is due to component $B \cos \theta$ or $B \cos \omega t$.

$$\therefore \quad \Phi = NAB \cos \omega t$$

$$\therefore \quad E = -\frac{d\Phi}{dt} = \omega NAB \sin \omega t = E_0 \sin \omega t,$$

where $E_0 = \omega NAB$.

Direction of e.m.f. Apply Fleming's right hand rule to the conductors through P and Q in Fig. 11.4b. The direction of motion of P and Q changes on passing the vertical. Hence the e.m.f. *reverses* at the vertical, or twice per revolution. The e.m.f. $E = E_0 \sin \omega t$ follows a sine wave pattern—it is a "sinusoidal" a.c.

Note that $E = 0$ when the coil is vertical and maximum flux then links the coil. Fig. 11.4c. This is because the rate of change is zero at this instant. E is a maximum when the coil is horizontal and no flux links the coil—the rate of change is then a maximum.

A.C. and D.C. For a.c., connect the ends of the coil to two *slip-rings*, which turn with the coil and press against two fixed brushes. For d.c., connect the ends of the coil to two insulated halves of a split-ring *commutator*, which turn with the coil and rotate against fixed brushes. The commutator reverses the connections to the external circuit joined to the brushes each time the coil passes the vertical. Thus two half-waves in the same direction are obtained. In this case,

$$\text{average e.m.f.} = \frac{2}{\pi} \times \omega NAB$$

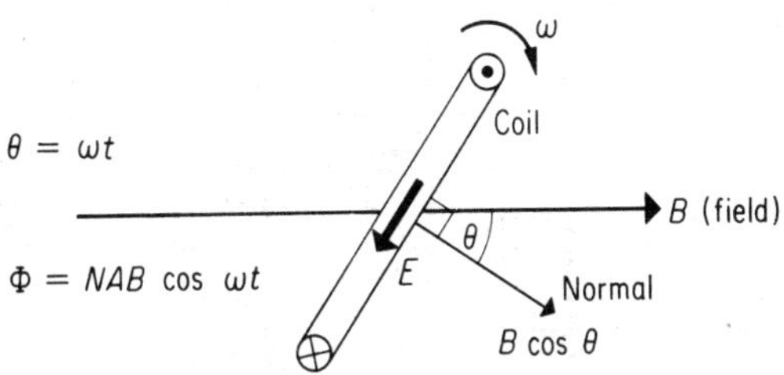

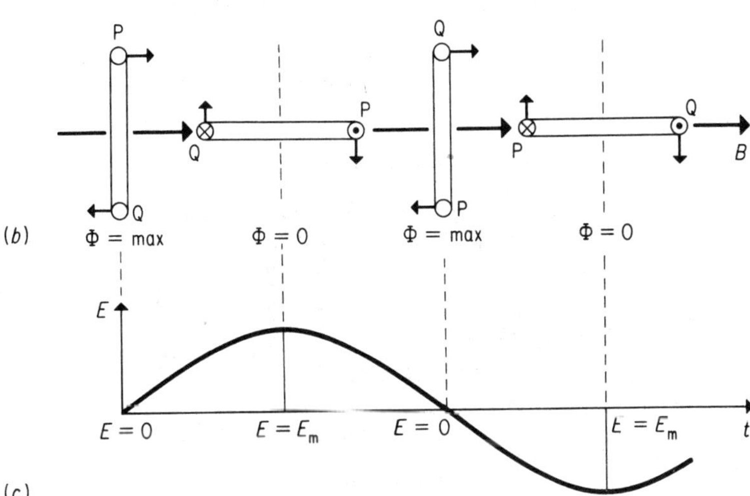

FIG. 11.4 Induced e.m.f. in simple dynamo.

Transformer. This changes alternating e.m.f. from low to high values, and vice versa. A commercial or soft ion transformer has (*i*) a primary coil, to which the voltage E_p is connected, (*ii*) a secondary coil, from which the new voltage E_s is obtained, (*iii*) a soft-iron laminated core on which both coils are wound. Since the flux linking a coil at any instant is proportional to the corresponding number of turns,

$$\frac{E_s}{E_p} = \frac{N_s}{N_p},$$

where N_s, N_p are the secondary and primary turns. Fig. 11.5a.

If no losses occurred, secondary current I_s/primary current I_p = N_p/N_s at every instant, since $I_s E_s = I_p E_p$ in this case.

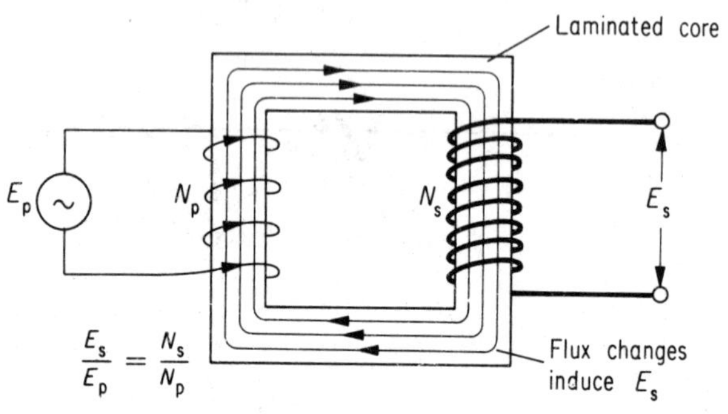

(a) Secondary open

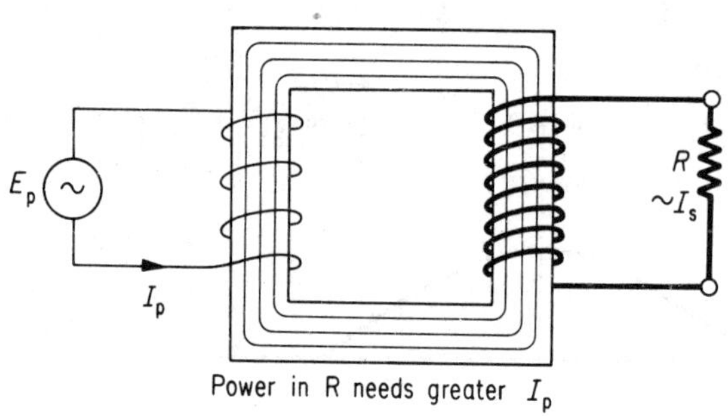

(b) Secondary closed

Fig. 11.5 Transformer.

When the secondary is on open circuit, only a small current flows in the primary. When the secondary is closed by a resistance, the primary current increases considerably. Fig. 11.5b. The back e.m.f. in the primary circuit is now reduced by the induced flux due to the secondary current flowing, which produces a bigger primary current. A higher primary current is necessary to provide the increased power now used in the secondary circuit.

Losses in transformers are: (*i*) copper or heat losses, (*ii*) eddy current losses, (*iii*) flux-leakage, (*iv*) hysteresis losses, the last three losses being in the core.

Flux-change and Q. When a *flux-change* Φ occurs in a circuit of total resistance R, a quantity of charge Q flows given by

$$Q = \frac{\Phi}{R}$$

Units: Q in coulomb (C), when Φ in weber (Wb) and R in ohms (Ω). Proof: At any instant, induced e.m.f. $E = -d\Phi/dt$. Hence current I at that instant $= E/R = -d\Phi/R \cdot dt$

$$\therefore \quad \int_0^t I \cdot dt = -\frac{1}{R} \int_{\Phi_1}^{\Phi_2} d\Phi,$$

where Φ_1 is the initial flux lining the circuit at $t = 0$ and Φ_2 is the final flux at t.

$$\therefore \quad Q = \frac{\Phi_1 - \Phi_2}{R} = \frac{\Phi}{R},$$

where $\Phi = \Phi_1 - \Phi_2 =$ flux change. Note that Q is independent of the time taken to make the flux change.

Measuring B. *Powerful magnet field.* Place a search coil (a small coil with a large number of turns) in the field so that all the flux enters the coil normally. Fig. 11.6a. Connect the coil to a ballistic galvanometer. Now withdraw the coil smartly from the field and observe the throw θ in the galvanometer. Then flux change $= NAB$, and so

$$Q = NAB/R = c\theta.$$

$$\therefore \quad B = \frac{Rc\theta}{NA}$$

R is the resistance of the coil plus the galvanometer; c is the galvanometer constant, found by discharging a capacitor of capacitance C having a p.d. V through the ballistic galvanometer, and observing the deflection θ_1, that is, $c = CV/\theta_1$. As N and A are known, B can be found.

Earth's field. The earth's field is weak—the horizontal component, B_H is about 2×10^{-5}T. A coil of large area A and a large number of turns N is hence needed to produce appreciable flux linkage Φ due to B_H—this coil is called an "earth inductor". With the *axis* of an earth inductor coil C pointing *east-west*, and the plane of C vertical, rotate the coil through $180°$. Fig. 11.6b. Then $\Phi =$ flux change $= NAB_H - (-NAB_H) = 2NAB_H$. Suppose R is the total resistance of coil and ballistic galvanometer connected to it. Then, if c is the constant of the galvanometer and θ the first throw, $Q = c\theta$.

$$\therefore \quad \frac{\Phi}{R} = c\theta$$

$$\therefore \quad \frac{2NAB_H}{R} = c\theta, \quad \text{or} \quad B_H = \frac{Rc\theta}{2NA} \quad . \quad . \quad . \quad . \quad (i)$$

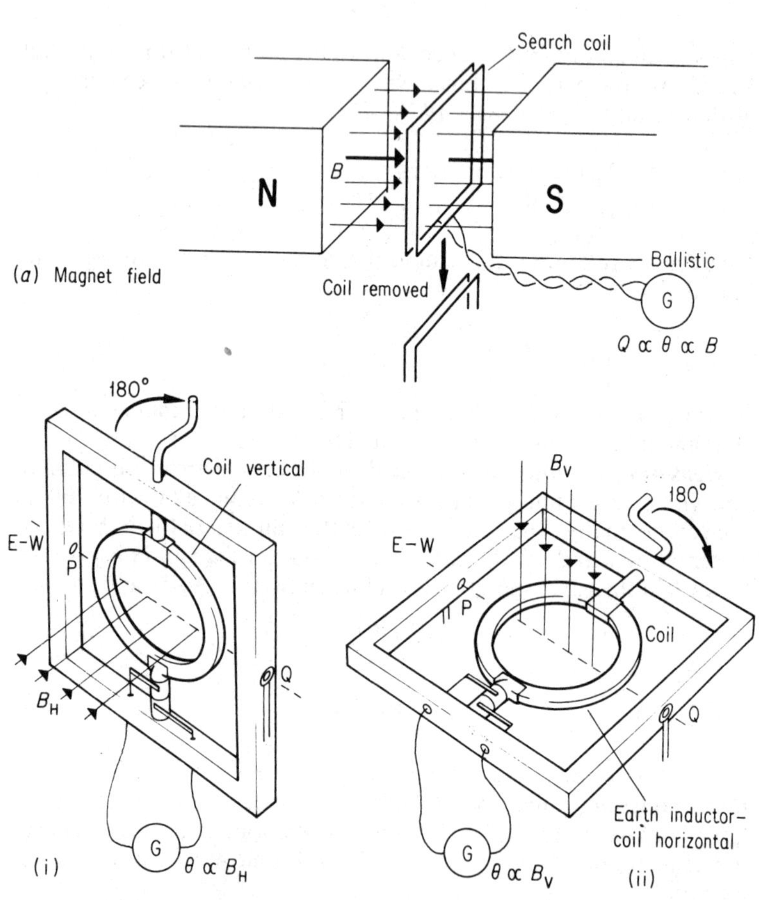

FIG. 11.6 Measuring B by flux change.

The vertical component, B_V, can be similarly found by rotating the coil from a *horizontal* position through 180°, and noting the throw of the ballistic galvanometer.

Angle of dip. The throws of the ballistic galvanometer for B_H and for B_V are observed. Suppose they are θ_1, θ_2 respectively. Then since $Q \propto \Phi$, and $\Phi = 2NAB_H$ and $2NAB_V$ respectively,

$$\frac{Q_2}{Q_1} = \frac{2NAB_V}{2NAB_H} = \frac{B_V}{B_H} = \frac{\theta_2}{\theta_1}$$

$$\therefore \quad \tan \delta = \frac{B_V}{B_H} = \frac{\theta_2}{\theta_1},$$

where δ is the angle of dip. Thus δ can be found from θ_2 and θ_1.

EXAMPLE

A flat coil of 150 turns, each of area 300 cm^2 and of total resistance 50 ohm, is connected to a circuit whose resistance is 40 ohm. Starting with its plane horizontal, the coil is rotated quickly through a half turn about a diametral axis pointing along the magnetic meridian. If the quantity of electricity which then flows round the circuit is 4 microcoulomb, find the intensity of the vertical component of the earth's magnetic field.

$$Q = \frac{\Phi}{R}.$$

Since $Q = 4/10^6$ coulomb, $R = 50 + 40 = 90$ ohm,

$$\therefore \quad \frac{4}{10^6} = \frac{\Phi}{90}$$

$$\therefore \quad \Phi = 36 \times 10^{-5} \text{ Wb}$$

Now flux change $\Phi = 2NAB_V = 2 \times 150 \times 300 \times 10^{-4}B_V$

$$\therefore \quad 2 \times 150 \times 300 \times 10^{-4}B_V = 36 \times 10^{-5}$$

$$\therefore \quad B_V = 4 \times 10^{-5} \text{ T.}$$

Self-inductance. The *self-inductance* (or *inductance*), L, of a coil may be defined in one of two ways:

(i) $L = E/(dI/dt)$, where E is the induced e.m.f. and dI/dt is the corresponding rate of change of current. Units: L in henry when E in volt, dI/dt in ampere per second.

(ii) Since $E = d\Phi/dt$ numerically, then, from (i), $L = d\Phi/dI$, or Φ/I if L is constant.

Thus L is the flux linkage per unit current. Units: L in henry (H) when Φ in weber (Wb) and I in ampere.

L for solenoid. $\Phi = NAB$, where $B = \mu NI/l$ (p. 105). Thus $\Phi = \mu N^2 AI/l$. Hence $\Phi/I = \mu N^2 A/l = L$

Units: L in henrys when μ (permeability) in H m^{-1}, A in metre^2, l in metre. For an air-core, $\mu = \mu_0 = 4\pi \times 10^{-7}$ H m^{-1}.

Energy stored in magnetic field of a coil carrying a steady current I_0 = work done against back e.m.f. while current rises to I_0 from zero

$$= \int_0^{I^0} E . I . dt = \int_0^{I^0} L\frac{dI}{dt} . I . dt = \tfrac{1}{2}LI_0{}^2.$$

When L is in H and I_0 is in A, the energy is in joule (J).

Growth of Current. When the circuit is made for an inductance (L) – resistance (R) d.c. series circuit, the current I at a time t is expressed by:

$$I = I_0(1-e^{-Rt/L}).$$

At the instant of switching on, the rate of rise of current is high and the back e.m.f. is high. The rate of rise of current decreases as the current grows, and the final steady value of current = I_0 = E/R, where E is the e.m.f. of the battery.

Mutual inductance. The mutual inductance, M, between two coils A, B is defined as the ratio $E_A/(dI_B/dt)$, where E_A is the induced e.m.f. in A when the rate of change of the current in B is dI_B/dt. Alternatively, $M = \Phi_A/I_B$, where Φ_A is the flux linking A when a current I_B flows in B.

Units: M in henrys when E_A in volt, I_B in ampere, t in second, or Φ_A in weber and I_B in ampere.

12. MAGNETIC FIELDS AND B.

Direction of field. The magnetic field B is always perpendicular to the current direction, and its direction (the movement of a north pole) can always be found from *Maxwell's corkscrew rule*. The "clock rule" (clockwise is south polarity) is useful for the coil or the solenoid cases.

Straight wire. The field lines or flux are circles concentric with the wire in planes perpendicular to the wire. Fig. 12.1*a*.

Narrow circular coil. The flux in a plane perpendicular to the coil passing through the middle is circular near the edges but straight and parallel (uniform) over a small region round the centre. Fig. 12.1*b*.

Helmholtz coils. These are two parallel equal circular coils placed a distance apart equal to r, their radius. Fig. 12.1*c*. The field on either side of the point midway between the coils is uniform over an appreciable distance, unlike the case of the single coil.

Solenoid. Inside the solenoid the lines are fairly straight and parallel, and perpendicular to the turns. Fig. 12.1*d*. The field B is then appreciably uniform in the middle. At the ends of the solenoid the field diminishes to a value $B/2$.

Outside the solenoid the flux pattern is similar to that round a bar magnet. *B* outside is low compared to the value in the middle of the solenoid.

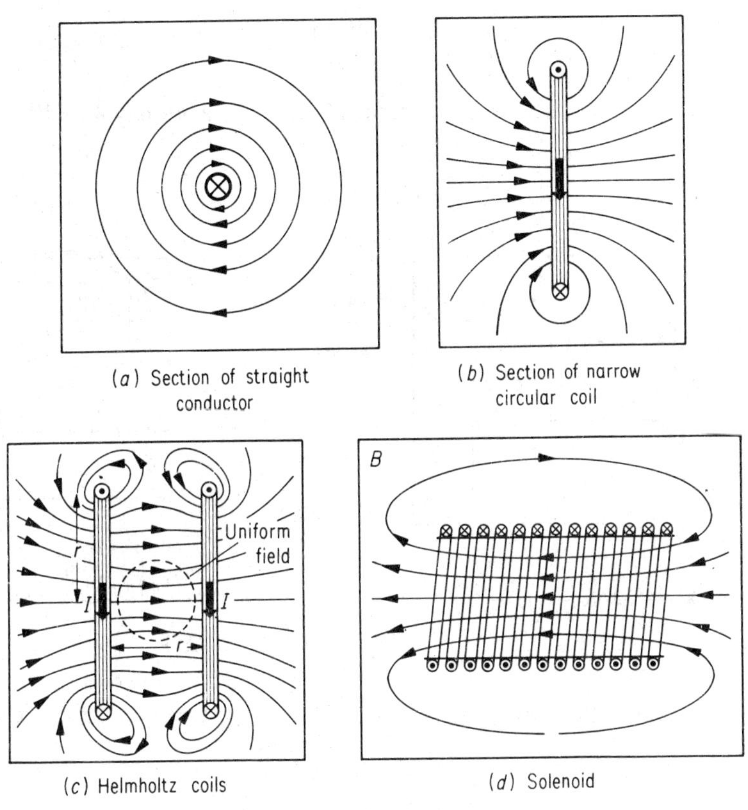

FIG. 12.1 Flux patterns round conductors.

Measurement of B. The flux density *B* can be measured by one of several methods:- (*i*) Hall probe, Fig. 12.2*a* (p. 102). (*ii*) Ballistic galvanometer (p. 97). (*iii*) Oscillating magnetic needle, Fig. 12.2*b* (p. 102). (*iv*) A.C. method and search coil—use mains frequency a.c., and arrange a small search coil C with a large number of turns so that the flux links it normally. Fig. 12.2*c*. The peak or maximum a.c. voltage induced in C is proportional to *B*. The voltage can be displayed on a sensitive cathode-ray oscillograph screen after amplification and measured. Hence a measure of *B* is obtained.

Biot-Savart law. For a current in a *small element* of a wire, the flux-density δB of the magnetic field at a point P outside the wire depends on four factors; it is proportional to (*i*) the current *I*, (*ii*) the sine of the angle θ between the element X and the line XP, (*iii*) the length δl of the element, and (*iv*) inversely proportional to the square of the distance XP, *r*. Thus

$$\delta B \propto \frac{I \delta l \sin \theta}{r^2}$$

In SI units, where μ_0 = permeability of vacuum $= 4\pi \times 10^{-7}$ H m^{-1},

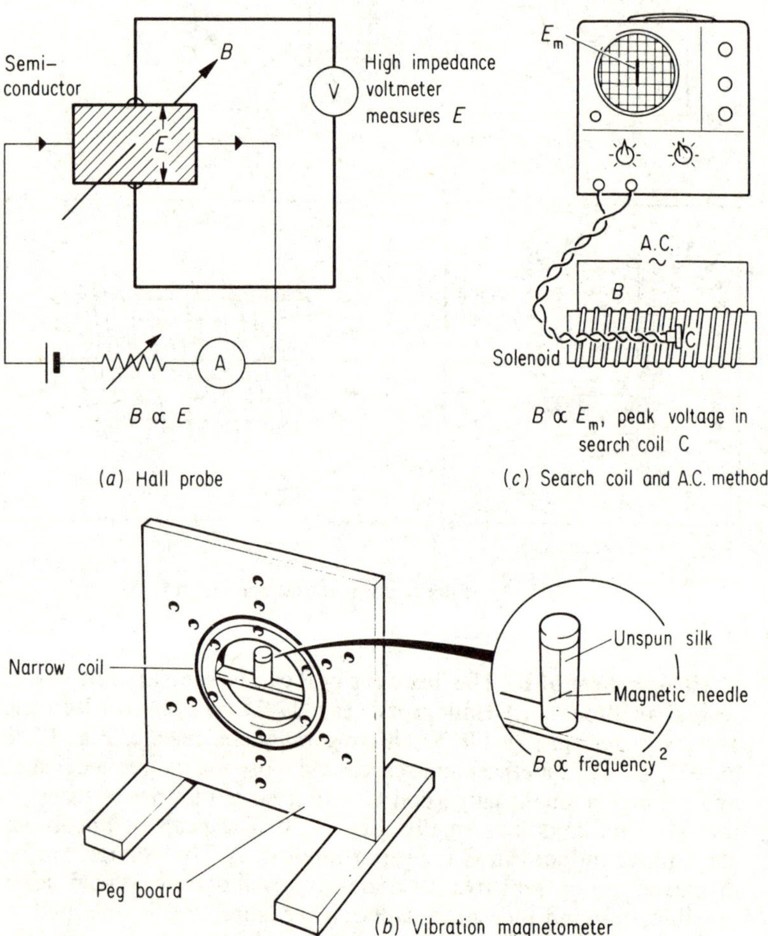

Fig. 12.2 Methods for measuring *B*

$$\delta B = \frac{\mu_0}{4\pi} \cdot \frac{I \delta l \sin \theta}{r^2}. \quad \text{(Fig. 13.3}a).$$

Units: δB in T when I in A, δl in m, r in m.

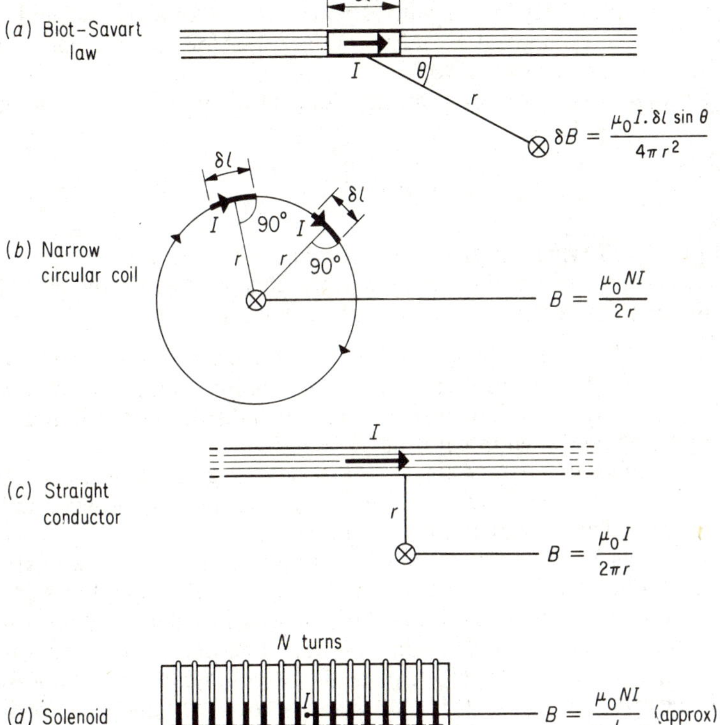

Fig. 12.3 Biot-Savart law and applications.

Narrow circular coil. At the centre of the current-carrying coil, Fig. 12.3b,

$$B = \int_0^l \frac{\mu_0}{4\pi} \cdot \frac{I \cdot dl \cdot \sin 90^\circ}{r^2} = \frac{\mu_0 I \sin 90^\circ}{4\pi r^2} \int_0^l dl$$

$$= \frac{\mu_0 Il}{4\pi r^2} = \frac{\mu_0 I \times 2\pi r N}{4\pi r^2} = \frac{\mu_0 NI}{2r} \quad \dots \dots \dots \quad (i)$$

Units: B in T when I in A, r in m, $\mu_0 = 4\pi \times 10^{-7}$ H m^{-1}.

(Note the absence of "π" in the expression for B—the flux pattern in the middle of the coil is straight and parallel.)

Investigation of B. Use a vertical board with pegs forming circles of different radii. See Fig. 12.2b. Wind wire round the pegs, pass current into the wire, and measure B by a Hall probe or other method. In this way one may investigate if $B \propto I$ for a given N and r, or if $B \propto 1/r$ for a given N and I, or if $B \propto 1/r^2$ for a *given length l* of wire and current I (see p. 102).

Straight wire. For an infinitely long straight wire, at a point distant r, Fig. 12.3c,

$$B = \frac{\mu_0 I}{2\pi r}.$$

Units: B in T when I in A, r in m, $\mu_0 = 4\pi \times 10^{-7}$ H m^{-1}.

Note the "π" in the denominator—this is due to the radial symmetry of the flux pattern round the wire (p. 101).

Proof: $B = 2\mu_0 I/4\pi \int_0^{\pi/2} I.\,dl.\,\sin\theta/a^2$, from Biot-Savart law, where a is the distance from δl to the point. Since $l = r \cot\theta$, $\delta l = -r \operatorname{cosec}^2\theta\,.\,\delta\theta$; also, $a = r \operatorname{cosec}\theta$. Substituting for dl and a, simplifying, and then integrating, $B = \mu_0 I/2\pi r$.

Investigation. To investigate if $B \propto I$ at a given distance r, or $B \propto 1/r$ for a given current I, use a long vertical straight wire, and a search coil and a.c. method, for example.

Resultant fields. The combined field pattern due to a downward current in a straight conductor in the earth's field, B_H, shows increased flux-density due west and diminished flux-density due east. As flux-density is a vector, the resultant flux-density due west is $(B+B_H)$; due east it is $(B-B_H)$ near the conductor and (B_H-B) further away; at the *neutral point* due east, $B = B_H$; and due north or due south, since B and B_H are perpendicular here, the resultant is $\sqrt{B^2 + B_H^2}$.

EXAMPLE

State and define *three* quantities necessary to specify completely the earth's magnetic field at a place. Illustrate your answer with a diagram. Describe briefly how you would determine *one* of these quantities.

A small bar magnet, free to oscillate in a horizontal plane about a vertical axis, is situated 8·0 cm magnetic west from a long straight vertical wire. When the magnet is set in oscillation it performs 25 oscillations in 20·0 seconds when no current flows in the wire, and the same number of oscillations in 15·0 seconds when a current flows down the wire. Calculate the current in the wire. [Horizontal component of earth's magnetic field $= 1\cdot8 \times 10^{-5}$ T] (*L*).

Suppose B_H and B are the respective flux-densities of the earth's and current's fields.

Then, in B_H, frequency $= \dfrac{25}{20} = \dfrac{5}{4} \, \text{s}^{-1}$

In total field, $B + B_H$, frequency $= \dfrac{25}{15} = \dfrac{5}{3} \, \text{s}^{-1}$

Since flux-density $\propto$ frequency^2,

$$\therefore \quad \frac{B + B_H}{B_H} = \frac{(5/3)^2}{(5/4)^2} = \frac{16}{9}$$

$$\therefore \quad B = \frac{7}{9}\, B_H = \frac{7}{9} \times 1{\cdot}8 \times 10^{-5} = 1{\cdot}4 \times 10^{-5} \text{ T}$$

For a long straight wire, $B = \mu_0 I / 2\pi r$, where $r = 8 \times 10^{-2}$ m

$$\therefore \quad \frac{4\pi \times 10^{-7} I}{2\pi \times 8 \times 10^{-2}} = 1{\cdot}4 \times 10^{-5}$$

$$\therefore \quad I = \frac{1{\cdot}4 \times 10^{-5} \times 4}{10^{-5}} = 5{\cdot}6 \text{ A.}$$

Solenoid. In the middle of a very long solenoid, with N turns and length l, Fig. 14.3d,

$$B = \frac{\mu_0 N I}{l} \quad \text{or} \quad \mu_0 n I \qquad \ldots \ldots \quad (i)$$

where n is the number of turns per unit length.

Units: B in T when I in A, l in m or n in m^{-1}.

At the ends of the solenoid, the flux-density has *half* the value in (i).

Proof: Take a length δx of the solenoid, so that the middle of the solenoid is a point on the *axis* of a narrow circular coil with $n \cdot \delta x$ turns, where n is the number of turns per metre. Using the formula for the value of flux-density on the axis of a circular coil of radius r, the flux-density B in the middle is given by $B = \mu_0(n \cdot dx)I r^2 / 2(r^2 + x^2)^{3/2}$.

For an infinitely long solenoid, the limits are $+\infty$ and $-\infty$. Substitute $x = r \tan \theta$, i.e. $\delta x = r \sec^2 \theta \cdot \delta\theta$. Then

$$B = \frac{\mu_0 n I}{2} \int_{-\pi/2}^{\pi/2} \frac{r^2 \cdot r \sec^2 \theta \cdot d\theta}{r^3 \sec^3 \theta} = \mu_0 n I$$

Investigation. Use a "Slinky" coil to vary the number of turns per metre, n, by pulling it out more—$n \propto 1/l$, where l is the length of the coil. Use a Hall probe, or search coil and a.c. method, to measure B in the middle of the coil. See Fig. 12.2c, p. 102.

105

Ampere's circuital law. Ampere showed that

$$\oint B \cdot dl = \mu_0 I,$$

where the integral $B \cdot dl$ represents the sum of all these quantities taken round a *closed path* and I is the total current enclosed by that path. Fig. 12.4.

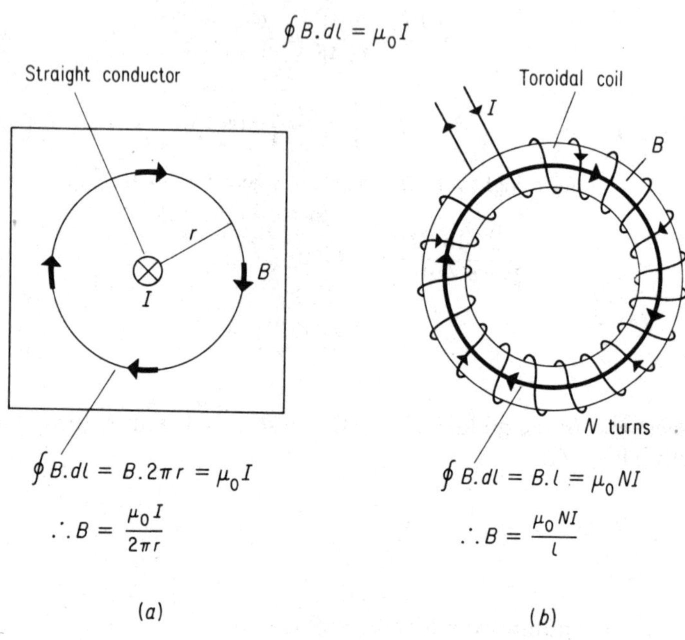

FIG. 12.4 Ampere's circuital law.

Investigation of law: Use a *Rogowski spiral* coil—a closed or toroidal coil, wound round rubber so that its shape can be changed easily, and with terminals in the centre for connection to a ballistic galvanometer so that the flux through it can be measured. (*i*) Place the closed coil in *any* position round a straight current-carrying wire and alter its shape or move it about. No deflection is observed on the galvanometer, showing that there is no flux change in the coil. (*ii*) Reverse the current in the conductor, observe the deflection and repeat this for different shapes of the coil. The deflection remains constant. (*iii*) Double the current and repeat—the deflection is then double.

Experiment shows that (*i*) $\int B \cdot dl$ is constant for a given current-carrying conductor, irrespective of the shape of the closed path, (*ii*) $\int B \cdot dl$ is proportional to I, the current.

Applications of Ampere's law. (*i*) *Straight conductor.* Fig. 12.4*a*. In this case, for a circular path of radius r,

$$\oint B \cdot dl = B \cdot 2\pi r = \mu_0 I$$

$$\therefore \quad B = \mu_0 I / 2\pi r$$

(2) *Toroidal coil.* Fig. 12.4*b*. For a circular path of length l of the coil,

$$\oint B \cdot dl = B \cdot l = \mu_0 N I,$$

since N coils each carry a current I. Thus

$$B = \mu_0 N I / l.$$

13. PROPERTIES OF MAGNETIC MATERIALS

Magnetic induction. (B) in a specimen is the flux-density in it. Fig. 13.1*a*. Unit: Telsa (T) or Wb m^{-2}.

Intensity of magnetization (M), also known as "magnetization", is

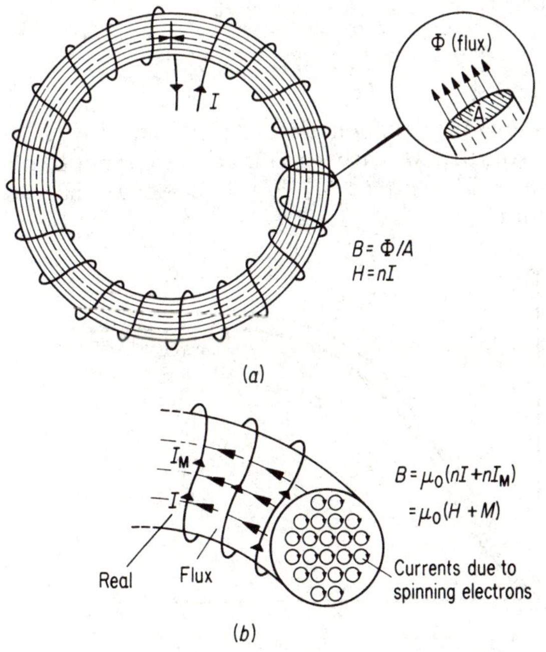

FIG. 13.1 Flux density in magnetic material.

the magnetic moment per unit volume of the specimen concerned, i.e. $M = m/V$ Unit: Am^{-1}.

Permeability (μ) is the ratio B/H. Unit: $H\ m^{-1}$. Incremental permeability $= \delta B/\delta H$. *Relative permeability*, $\mu_r = \mu/\mu_0$, where μ_0 is the permeability of a vacuum. Unlike μ, μ_r has no units.

Magnetizing field intensity (H) is nI, where n is the number of turns per metre in a solenoid and I is a current. Unit: $A\ m^{-1}$.

Susceptibility (χ_m) is the ratio M/H. It has no units, being a ratio.

Hysteresis is the name given to the lagging of B behind H when a specimen is taken through a magnetic cycle.

Remanence (B_r). This is the flux density remaining in a specimen when the magnetizing field is reduced to zero.

Coercive force (H_c). This is the field intensity required to reduce the flux-density in the specimen to zero.

General relations.
$$B = B_0 + B_M, \qquad \qquad (1)$$

where B_0 is the flux-density due to the magnetizing current flowing and B_M is that due to the magnetization of the material. Since $B_0 = \mu_0 H$ and $B_M = \mu_0 M$, Fig. 13.1b,

$$B = \mu_0 (H + M) \qquad \qquad (2)$$

Dividing by H,
$$\mu = \mu_0 (1 + \chi_m)$$

$$\therefore \quad \frac{\mu}{\mu_0} = \mu_r = 1 + \chi_m \qquad \qquad (3)$$

Fig. 13.2a shows the variation of B and of M with H as H is increased from zero. Note that M reaches a saturation value but B continues to increase, since $B = \mu_0(H + M)$. Fig. 13.2b shows the variation of μ and χ_m with H.

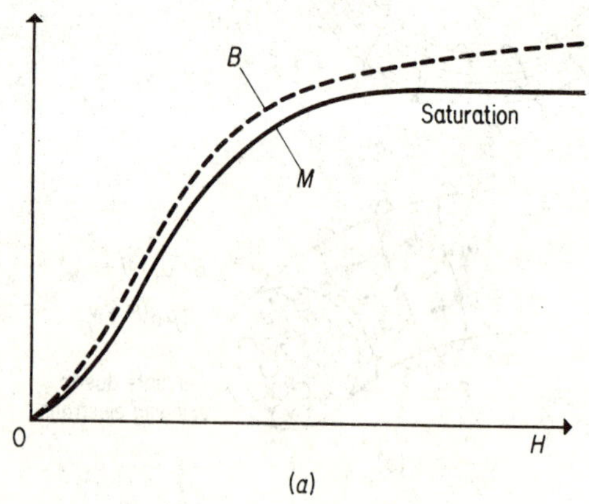

FIG. 13.2. Variation of B and M with H.

108

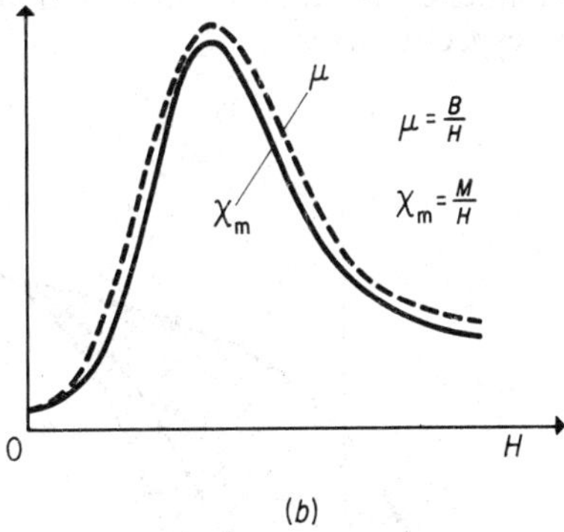

Fig. 13.2 Variation of μ and χ_m with H

EXAMPLE

A long iron specimen is placed inside a long solenoid which has 1000 turns per metre. When a current of 1 ampere is passed, the relative permeability of the material is then 700. Calculate the intensity of magnetization M and the flux-density B in the iron.

Magnetizing field intensity $H = nI = 1000 \times 1 = 1000$ A m^{-1}

$$\therefore \quad B = \mu H = \mu_r \mu_0 H = 700 \times 4\pi \times 10^{-7} \times 1000$$

$$= 0{\cdot}88 \text{ Wb m}^{-2} \quad \ldots \ldots \ldots \ldots \ldots \ldots \quad (i)$$

From $$B = \mu_0(H + M),$$

$$\therefore \quad M = \frac{B}{\mu_0} - H$$

Now $$\frac{B}{\mu_0} = \mu_r H = 700 \times 1000$$

$$\therefore \quad M = 700 \times 1000 - 1000 = 699\,000 \text{ A m}^{-1} \quad \ldots \ldots \quad (ii)$$

B-H curve. When a magnetic material is taken through a magnetic cycle of H and B is measured, typical results are shown in Fig. 13.3.

The B—H loop is a closed curve. As H is increased from zero initially, B increases continuously although the material becomes magnetically saturated, since $B = \mu_0(H + M)$. Thus the top part of the B—H curve is *not* horizontal. The energy lost due to hysteresis $= \int B.dH$, and is thus proportional to the area of the loop. Steel has a wide hysteresis loop; soft iron has a relatively narrow loop, showing that the hysteresis loss is smaller (Fig. 13.4). The coercive force (H_c) and the remanence (B_r) are the values of H and B where the B—H loop cuts the respective axes.

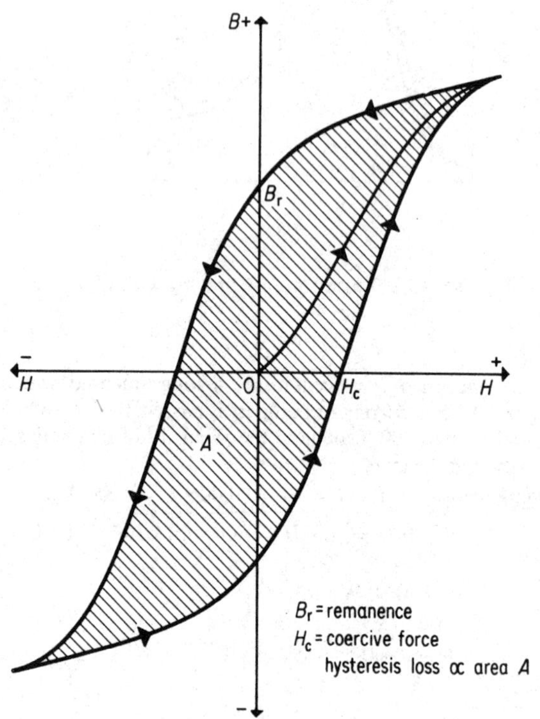

FIG. 13.3 *B – H* curve.

Properties of magnets and transformer cores.   (1) A *permanent magnet* material should have a high value of remanence and of coercive force, and a high susceptibility.

(2) An *electromagnet* material should have a high susceptibility at low magnetizing fields, and a low value of remanence and of coercive force.

(3) A *transformer core* material should have a high permeability at low magnetizing fields, low hysteresis loss, and high resistivity to reduce eddy-current losses.

(4) A *telephone diaphragm* material should have a high incremental permeability, and low hysteresis and eddy-current losses. (The attractive force F on the diaphragm is proportional to B^2, and hence

the change $\delta F \propto 2B.\delta B$; the permanent magnet in the earpiece
gives a high value of B, and a diaphragm of high incremental
permeability gives a high value of δB.)

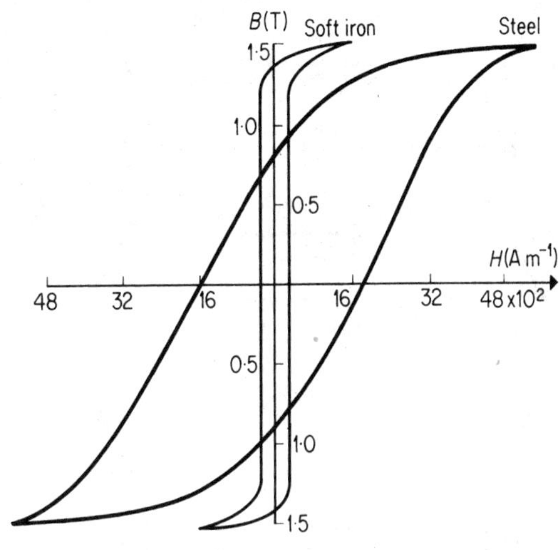

FIG. 13.4 $B - H$ curves for steel and soft iron.

DIAMAGNETISM, PARAMAGNETISM, FERROMAGNETISM

Diamagnetic materials   settle at right angles to a non-uniform
magnetic field. Fig. 13.5a. The magnetism induced is weak and
opposite to that which would be induced if iron or steel were placed
in the field. The susceptibility is thus low and negative. From
$\mu = 1 + \chi_m$, the permeability is slightly less than 1. The susceptibility
is independent of temperature. Bismuth and quartz are diamagnetic.

111

Paramagnetic materials are feebly magnetized when situated in a magnetic field. Fig. 13.5*b*. The susceptibility is thus very small, and permeability slightly greater than 1. The susceptibility decreases as the temperature increases. Copper and platinum are paramagnetic.

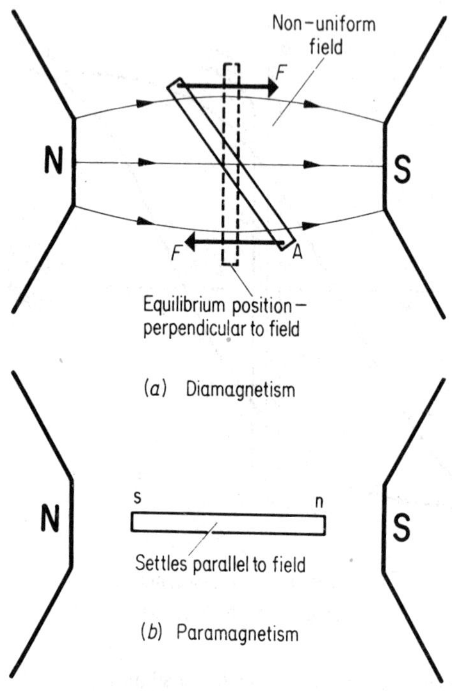

FIG. 13.5 Diamagnetism. Paramagnetism.

Ferromagnetic materials are those which are strongly magnetized in a magnetic field. They have small regions or *domains* inside them which are powerfully magnetized. See Fig. 13.6 *a, b, c.* Iron, steel, nickel, cobalt and their alloys are ferromagnetic materials. Above a certain temperature known as the Curie point, about 750°C for iron, ferromagnetic materials become paramagnetic.

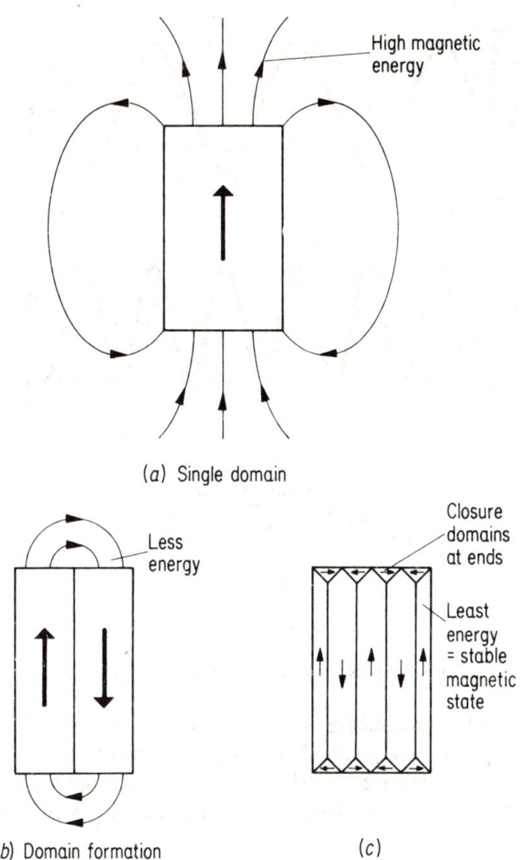

FIG. 13.6 Domains in ferromagnetic materials.

14. A.C. CIRCUITS. ELECTROMAGNETIC WAVES

Root-mean-square (r.m.s.) value.   The root-mean-square (r.m.s.) value of a current is that value of steady (direct) current which gives the same mean heating effect. Thus the r.m.s. value is the square-root of the average of all the I^2 values taken over one cycle. Fig. 14.1a, b.

For a sine wave,

$$\text{r.m.s. value} = \frac{1}{\sqrt{2}} \times \text{peak value} = 0.71 \times \text{peak value}.$$

$$\text{Average value} = \frac{2}{\pi} \times \text{peak value} = 0.64 \times \text{peak value},$$

the average being calculated over half a cycle. See also p. 95.

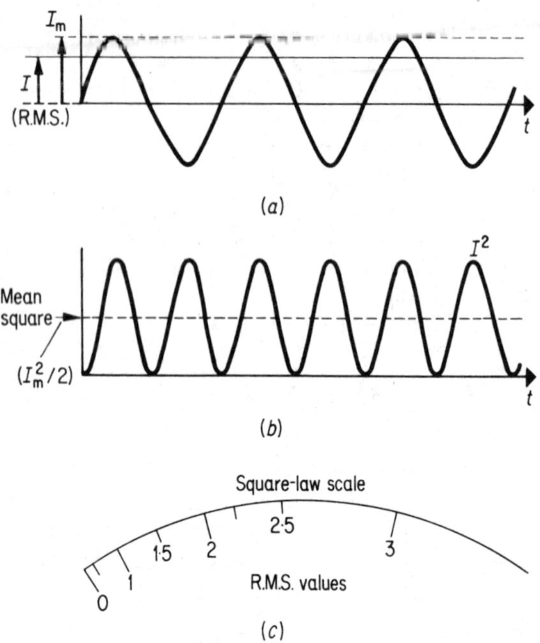

FIG. 14.1 Root-mean-square value of a.c.

Instruments for measuring A.C.   (1) Hot-wire ammeter (p. 72)— this has a square-law scale, Fig. 14.1c, (2) moving-iron ammeter, (3) moving-coil instrument with rectifier (such as selenium rectifier). (4) The cathode-ray oscillograph measures the peak (maximum) value.

The r.m.s. value of a current can be measured by passing it into a coil in a calorimeter with water, and observing the temperature rise θ in a time t. The heat, H, is then obtained and the current I (r.m.s.) is calculated from $I^2 Rt = H$, where R is the coil resistance.

A.C. circuits.   *Pure resistance, R.* $I = V/R$, where I, V are r.m.s. values. The alternating current and voltage are in phase. Fig. 14.2a.

Pure inductance, L. $I = V/X_L$, where $X_L = 2\pi f L$ = reactance of coil. The voltage leads on the current by 90°, Fig. 14.2b. X_L is in ohms when *L* is in *henrys* and *f* in Hz. The phase difference can be demonstrated with a double-beam cathode-ray oscillograph, by superimposing the wave-form of the voltage across the coil on the wave-form of the voltage across a resistance in the a.c. circuit, since the latter is in phase with the current flowing.

Pure capacitance, C. $I = V/X_C$, where $X_C = 1/2\pi f C$ = reactance of capacitor. The voltage lags on the current by 90°. Fig. 14.2c. X_C is in ohms when C is in *farads* and *f* in Hz. The phase change can be demonstrated with the C.R.O., as stated above.

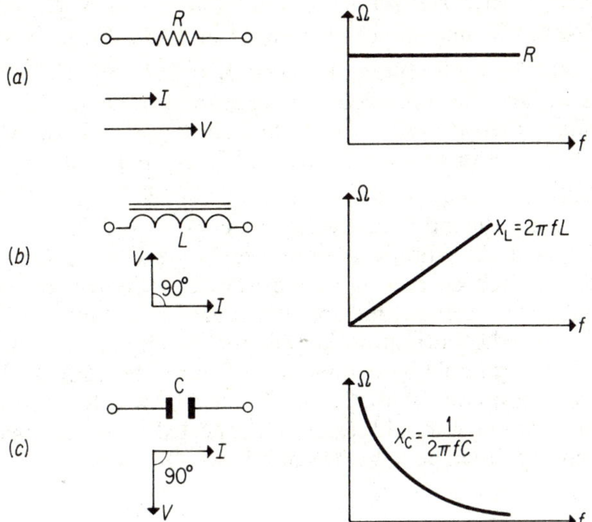

FIG. 14.2 *R* and *L* and *C* in a.c. circuit.

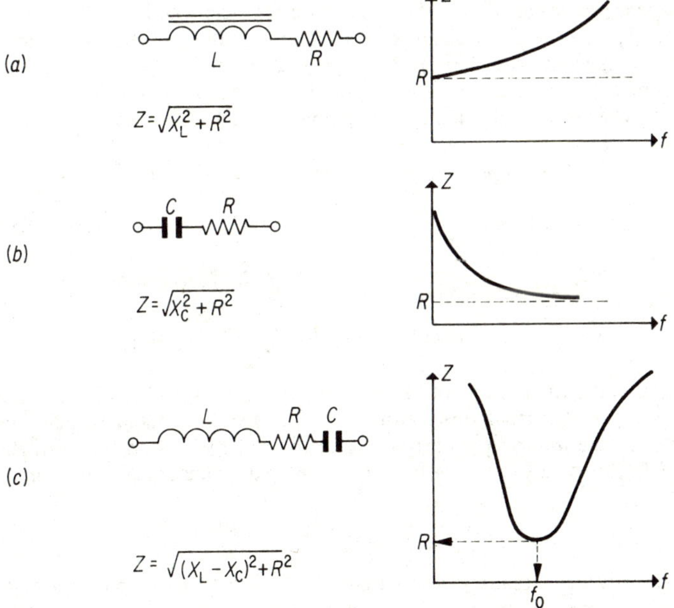

FIG. 14.3 Series (*a*) *L*, *R* (*b*) *C*, *R* (*c*) *L*, *C*, *R*.

Inductance L and resistance R in series. $I = V/Z$, where $Z = \sqrt{R^2 + X_L^2}$ = impedance of circuit. Fig. 14.3a. I lags on V by $\tan^{-1}(X_L/R)$. The total voltage V across L and $R = \sqrt{V_L^2 + V_L^2}$, where V_L, V_R are the individual voltages across L and R.

Capacitance C and resistance R in series. $I = V/Z$, where $Z = \sqrt{R^2 + X^2}$ = impedance of circuit. Fig. 14.3b. I leads on V by $\tan^{-1}(X_C/R)$. The total voltage V across C and $R = \sqrt{V_C^2 + V_R^2}$ where V_C, V_R are the individual voltages across C and R.

Note. When a d.c. supply is connected to a coil of wire the "opposition" to the current is the resistance R. When an alternating voltage is connecting to the same coil, however, the "opposition" to the current is largely the reactance X_L, as the resistance R is negligible by comparison. The reactance X_L will be large if a soft-iron core is used, and the current will then be low. A transformer primary should never be connected to the d.c. mains, as the current may then be so high as to burn out the coil.

EXAMPLE

Explain what is meant by the *root mean square* of an alternating potential difference and why its value is used as a measure of the alternating potential difference. Show how the r.m.s. value and the peak value of a sinusoidal potential difference are related.

A moving iron voltmeter, calibrated to read unidirectional potential differences, consists of a coil of resistance 600 ohms and of inductance 1·25 henry in series with a resistance of 4400 ohms. What will the voltmeter read when connected to an alternating supply of 250 volts at a frequency of 200 Hz? Comment on the answer by indicating the limitation of a moving iron voltmeter when measuring an alternating potential difference. (*L*)

$$\text{Impedance of whole circuit, } Z = \sqrt{X_L^2 + R^2}$$

$$= \sqrt{(2\pi \cdot 200 \cdot 1\cdot25)^2 + 5000^2} \; \Omega$$

$$= 5{,}240 \; \Omega \text{ (approx.)}$$

$$\therefore \quad \text{current flowing, } I = \frac{V}{Z} = \frac{250}{5{,}240} = \frac{1}{21} \text{ A.}$$

This is the r.m.s. value of the current. It is the magnitude of the steady or unidirectional current which would produce the same magnetic effect in the meter as the alternating current. Since the *resistance R* in the voltmeter is 5000 Ω, this steady current would produce a p.d. or meter reading given by

$$V = IR = \frac{1}{21} \times 5000 = 238 \text{ V}$$

The voltmeter reading is less than 250 V by 12 V, an error of about 5%. At higher frequencies the reactance of the coil increases further relative to the 5000 Ω, leading to greater error.

Series L, C, R circuit. $I = V/Z$, where $Z = \sqrt{R^2 + (X_L - X_C)^2}$. I lags (or leads) on V by $\tan^{-1}(X_L \sim X_C)/R$. The current (I)–frequency (f) graph has a peak (resonance) value.

At resonance,

(i) the impedance Z reduces to a pure resistance R, Fig. 15.3c,

(ii) the current (a maximum) $= V/R$,

(iii) the resonant frequency, f_0, $= 1/2\pi\sqrt{LC}$ $(X_L = X_C)$,

(iv) the voltage $V_L =$ the voltage $V_C = Q \times V$, the applied voltage, where $Q = 2\pi f_0 L/R = (1/R) \times \sqrt{L/C}$.

(v) the voltage across both L and C is zero, since V_L is 180° out of phase with V_C.

Parallel A.C. circuits. The *vector sum* of the *currents* in the parallel branches is equal to the current in the main circuit.

A coil (L, R) in parallel with a capacitor (C) has a maximum impedance value Z at resonance given by $Z = L/CR$. The resonant frequency is given by $f_0 = 1/2\pi\sqrt{LC}$.

Power in A.C. circuits. Power is absorbed only in the resistance, R. No power is absorbed in a pure inductance L or a pure capacitance C. Fig. 14.4a. Thus, generally, power absorbed $P = I^2R$, where I is in amperes r.m.s., R in ohms and P in watts; or V^2/R, or IV, where I, V are both in r.m.s.

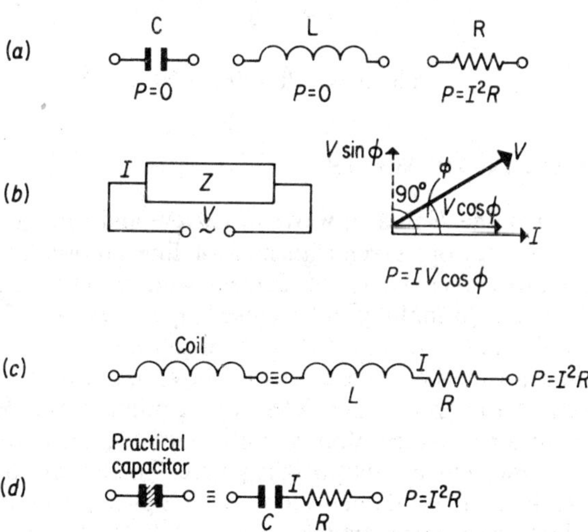

Fig. 14.4 Power in a.c. circuit.

117

Power factor (P.F.) is defined as the ratio: *power absorbed/IV*, where I is the main current and V is the voltage applied, in r.m.s. Thus power factor = true power absorbed/apparent power supplied. In any a.c. circuit, if φ is the phase angle, Fig. 14.4*b*.

$$\text{true power absorbed} = IV \cos \varphi \quad \ldots \quad (i)$$

$$\therefore \quad \text{power factor} = \cos \varphi \quad \ldots \quad (ii)$$

In the case of a *coil* having inductance L and resistance R, tan φ = X_L/R (p.116). Hence the power factor of a coil = $\cos \varphi$ = $R/\sqrt{R^2 + X_L^2}$. The power factor is low when X_L is large compared with R, in which case the power absorbed (only in R) is low. Fig. 14.4*c*. In the case of capacitors, an air capacitor has practically a zero power factor, a mica capacitor has a low power factor, and an electrolytic capacitor has a high power factor. Fig. 14.4*d*.

EXAMPLE

A coil of 0·01 H and resistance 20 Ω is joined in series with a 0·04 μF capacitor. An a.c. voltage of 10 V is connected across the whole arrangement at the resonant frequency f_0. Calculate f_0 and the power absorbed in the circuit.

$$(i) \quad f_0 = \frac{1}{2\pi\sqrt{LC}} = \frac{1}{2\pi\sqrt{0\cdot01 \times 0\cdot04 \times 10^{-6}}} = 8\,000 \text{ Hz (approx.)}.$$

(ii)   At resonant frequency, the reactances cancel and 20 Ω resistance is left.

$$\therefore \quad \text{current, } I = \frac{V}{R} = \frac{10}{20} = 0\cdot5 \text{ A}$$

$$\therefore \quad \text{power} = I^2R = 0\cdot5^2 \times 20 = 5 \text{ W}.$$

ELECTROMAGNETIC WAVES

Transmission line. Standing waves of current and voltage.   At the resonant frequency of a given transmission line, an oscillator sets up standing (stationary) waves of current and voltage. Fig. 14.5*a*. Fig. 14.5*b* shows the analogy with a closed pipe at resonant frequency. Here air molecules move and a node of displacement is at the closed end. In the transmission line, electrons move and a node of current is obtained at the ends, where electrons cannot move. Voltage is analogous to air pressure. Voltage antinodes may be detected by a small neon tube, which glows brightly here. Current antinodes may be detected by a small filament lamp, which glows brightly here.

Propagation of electromagnetic waves.   A finite magnetic field B, moving normally to its direction with a velocity v, sets up an electric field E given by $E = vB$. B, E and v are mutually at right angles. Fig. 14.5*c*. An electric field E, moving with velocity v, sets up a

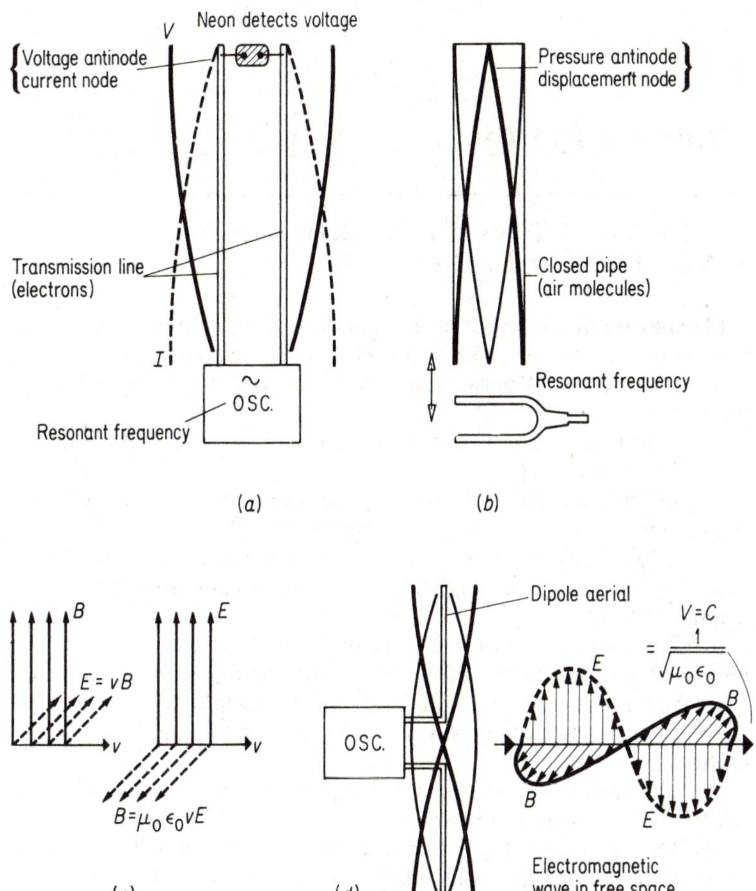

FIG. 14.5 Standing waves of I and V. Electromagnetic waves.

magnetic field B given by $B = \mu_0 \varepsilon_0 vE$. The fields E and B will be self-sustaining if $\mu_0 \varepsilon_0 v = 1/v$, or $v = 1/\sqrt{\mu_0 \varepsilon_0}$. Since $\mu_0 = 4\pi \times 10^{-7}$, $\varepsilon_0 = 8\cdot854 \times 10^{-12}$, then $v = 3 \times 10^8$ m s^{-1} (approx.). This is the speed c of light waves. Electromagnetic waves may be propagated in space by aerials. Fig. 14.5d. They consist of electric E and magnetic fields B moving with a speed v of 3×10^8 m s^{-1}; E and B act in planes perpendicular to v so that electromagnetic waves are transverse waves.

15. CHARGES IN ELECTRIC AND MAGNETIC FIELDS

Phenomena of discharge tube. The following changes occur in a gas under high voltage as the pressure, p, is lowered:—

(i) Streamers pass between anode A and cathode C ($p = 20$ mm mercury).

(ii) Luminous broad (positive) column between A and C ($p = 10$ mm mercury).

(iii) Negative glow on C, dark space (Faraday's dark space), and then positive column to A ($p = 5$ mm mercury).

(iv) Glow moves away from C, leaving Crookes' dark space near C, the positive column shortens and breaks into discs ($p = 0.5$ mm of mercury).

(v) The positive column moves back to A and disappears, the Crookes' dark space fills all the tube, and the discharge tube glows ($p = 0.01$ mm mercury). *Cathode rays now present inside tube.*

Experiments on cathode rays. Rays appear to come from C, the cathode.

(1) The rays travel in straight lines—shown by sharp shadow of Maltese cross forming the anode.

(2) The rays are deflected by a magnetic field—shown by movement of luminous glow when magnet brought near.

(3) The rays cause fluorescence—zinc sulphide has a greenish-blue colour, and glass a greenish colour.

(4) The rays carry a quantity of negative electricity—shown by deflecting the rays in a Perrin tube into an electrometer.

(5) The rays cause heat—shown by white-hot platinum when latter placed in path of rays.

(6) The rays produce X-rays when impinging on certain metals—shown by the X-ray tube (p. 152).

Nature of cathode rays SIR J. J. THOMSON measured the ratio e/m ("specific charge") for cathode rays, where e is the charge and m is the mass. He found that m was about 1/2000 of the mass of a hydrogen atom (see p. 129). The sign of the charge can be deduced by applying Fleming's left-hand rule to the deflection in experiment (2) above, and the charge was found to be negative. *Cathode rays are thus tiny particles carrying a negative charge* and are now called *electrons.* The charge on an electron, e, is numerically about 1.6×10^{-19} C.

Electric fields. Current and electron drift. The current I in a metal is proportional to the electron drift velocity v. Fig. 15.1a, The number of electrons crossing a section of area A per second $= A.v.n$, where n is the number per unit volume. Thus $I = nevA$. The current density $j = I/A = nev$.

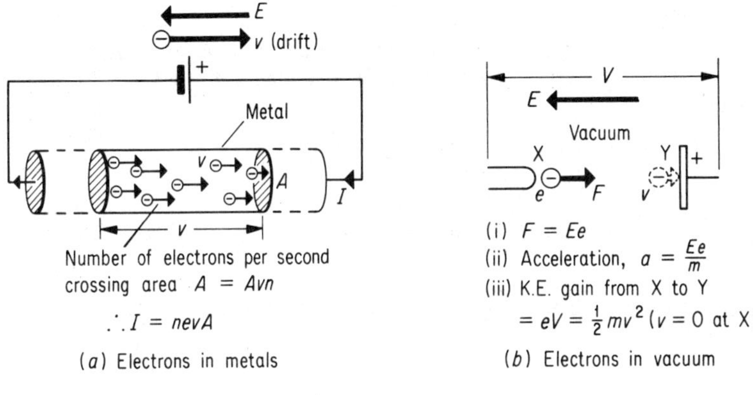

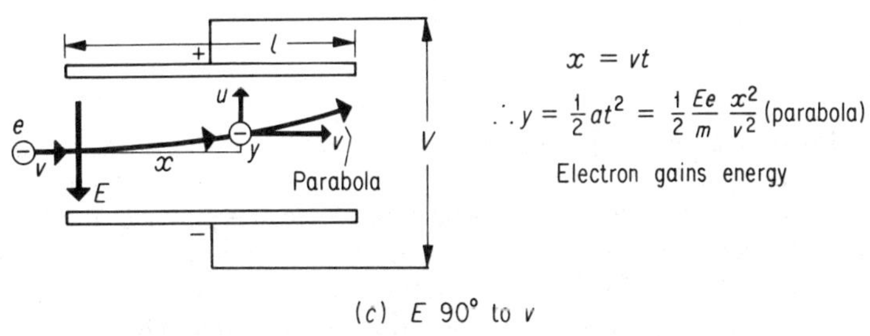

FIG. 15.1 Electron motion in conductor and vacuum.

If a current of 2 A flows in a copper wire of cross-section area $1\ \text{mm}^2$ or $10^{-6}\ \text{m}^2$, and $n = 8.10^{28}\ \text{m}^{-3}$, then, since $e = 1.6 \times 10^{-19}$ C,

$$v = I/neA = 2/(8.10^{28} \times 1.6 \times 10^{-19} \times 10^{-6})$$
$$= 1.6 \times 10^{-4}\ \text{m s}^{-1}$$

Motion in vacuum. Fig. 15.1b shows an electron liberated at a cathode X with a velocity $u = 0$. Suppose a p.d. V is applied between X and plate Y. Then if E is the electric intensity between X, Y,

$$\text{force } F \text{ on electron} = Ee \quad . \quad . \quad . \quad . \quad . \quad (1)$$

Units: E in V m^{-1}, e in C, then F in N. If m is the mass of an electron, then

$$\text{acceleration}, \ a = \frac{Ee}{m} \qquad \ldots \ldots \ldots \quad (2)$$

Units: E in V m^{-1}, e in C, m in kg, then a in m s^{-2}.

If v is the velocity on reaching Y, then, since energy gained $= eV$,

$$eV = \tfrac{1}{2}mv^2 \qquad \ldots \ldots \ldots \ldots \quad (3)$$

EXAMPLE

In Fig. 13.1b, XY $= 1$ cm, $V = 2000$ V. Find (i) the force on an electron, (ii) its acceleration, (iii) the velocity v at Y, assuming $e/m = 1\cdot8 \times 10^{11}$ C kg^{-1} and $e = 1\cdot6 \times 10^{-19}$ C.

(i)
$$E = \frac{2000 \text{ V}}{1 \times 10^{-2} \text{ m}} = 2 \times 10^5 \text{ V m}^{-1}$$

$$\therefore \quad F = Ee = 2 \times 10^5 \times 1\cdot6 \times 10^{-19} = 3\cdot2 \times 10^{-14} \text{ N}$$

(ii) Acceleration, a, $= \dfrac{Ee}{m} = 2 \times 10^5 \times 1\cdot8 \times 10^{11} = 3\cdot6 \times 10^{16}$ m s^{-2}

(iii)
$$\tfrac{1}{2}mv^2 = eV$$

$$\therefore \quad v = \sqrt{\frac{2e}{m}V} = \sqrt{2 \times 1\cdot8 \times 10^{11} \times 2000}$$

$$= 2\cdot7 \times 10^7 \text{ m s}^{-1}.$$

Particle path. Suppose an electron, moving with uniform velocity v enters a uniform electric field E between two parallel horizontal plates of length l. Fig. 15.1c. The force F on the particle due to E is always vertical. The *horizontal* motion is therefore unaffected. Hence time t to travel a horizontal distance $x = x/v$.

For vertical motion, acceleration $a = Ee/m$. Hence vertical displacement y in time $t = \tfrac{1}{2}at^2$ (vertical velocity is zero initially)

$$\therefore \quad y = \tfrac{1}{2}\frac{Ee}{m} \cdot \frac{x^2}{v^2} = \frac{Ee}{2mv^2} \cdot x^2 \qquad \ldots \ldots \quad (4)$$

The electron path is thus a parabola. When the particle reaches the ends of the plates, $x = l$. Hence vertical displacement on leaving plates, $d = Eel^2/2mv^2$.

Magnetic field. Force F on a charge q moving normally to a *magnetic field* B with velocity v is given by $F = Bqv$. Fig. 15.2a.

Suppose an electron, moving with velocity v, enters a *perpendicular* uniform field B. Fig. 15.2b. The force $F = Bev$ and is always perpendicular to the motion of the particle. It is therefore a deflecting

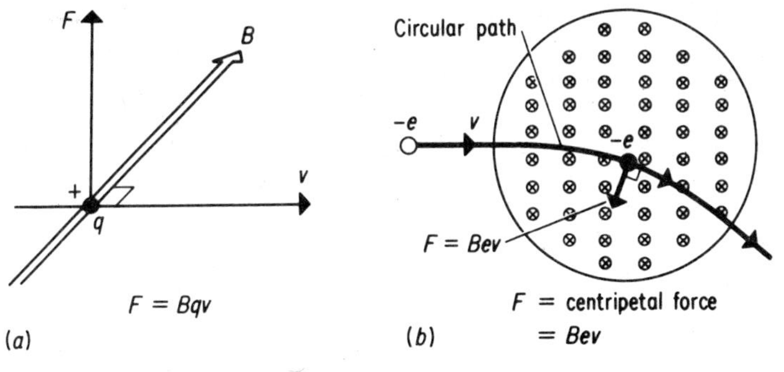

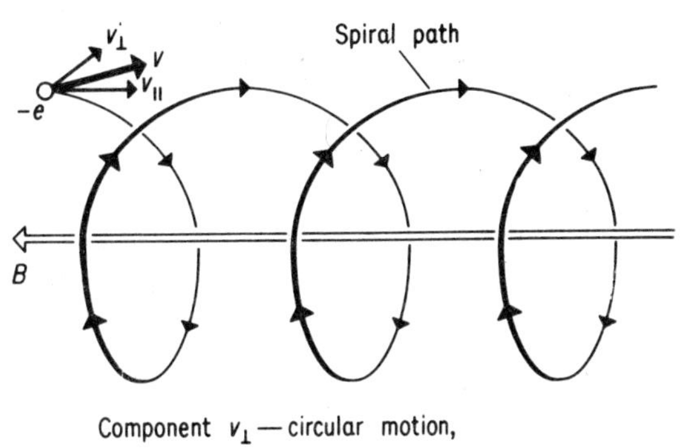

Fig. 15.2 Electron motion in magnetic fields.

force, unlike the force Ee in Fig. 15.1a; the electron gains no energy from the field since $F = Bev$ is always perpendicular to its motion. Further, F is constant. It is hence a constant *centripetal force*. Thus the electron is deflected in a circular path of radius r, where

$$F = Bev = \frac{mv^2}{r}$$

or

$$\therefore r = \frac{mv}{Be} \quad \ldots \ldots \ldots \quad (5)$$

From (5), the radius r is proportional to the *momentum* of the particle.

Strong magnetic field. If the field B is strong, an electron beam can be deflected into a complete circular path. See p. 128.

Helical path. If an electron beam enters a magnetic field B at an angle α to B less than 90°, the beam is deflected along a spiral or helical path. Reason: The component velocity $v \sin \alpha$ perpendicular to B produces circular motion; the component $v \cos \alpha$ parallel to B is unaffected by B and produces translational motion. Fig. 16.5c. The motion is thus helical.

The "pitch" p, of the spiral $= v \cos \alpha . T$, where T is the time for one complete revolution in circular motion. Now $T = 2\pi r/v \sin \alpha = 2\pi mv \sin \alpha/Bev \sin \alpha$, from (4) $= 2\pi m/Be$. Hence $p = 2\pi mv \cos \alpha/Be$.

Perpendicular Magnetic and Electric fields.   Suppose an electron, moving with velocity v, enters uniform perpendicular fields B and E. Fig. 15.3a. The force $F = Bev$ is then *parallel* to the force $F = Ee$.

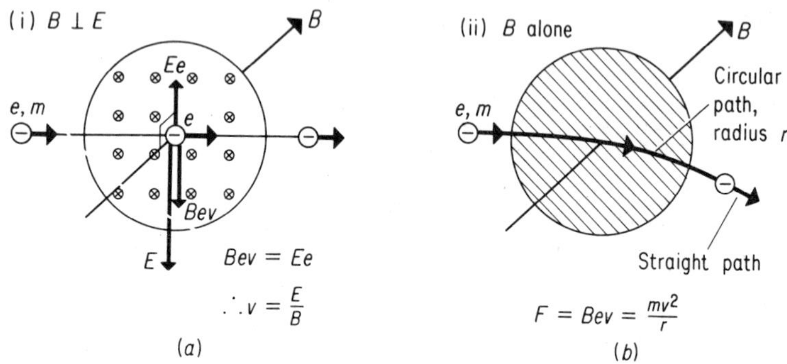

FIG. 15.3 Electrons in perpendicular fields of B and E.

If the forces are equal and opposite, there is no resultant deflection of the moving charge. This occurs when $Bev = Ee$, so that $v = E/B$ (p. 127). With B alone, the path is circular. Fig. 15.3b.

Parallel Magnetic and Electric fields. Suppose a beam of positive ions, each of charge q and mass m moving with uniform velocity v, enters the fields. Fig. 15.4. The forces Bqv and Eq are perpendicular,

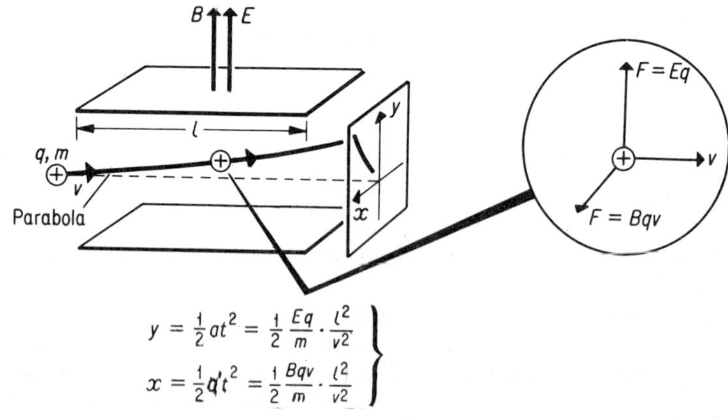

$$y = \tfrac{1}{2}at^2 = \tfrac{1}{2}\frac{Eq}{m} \cdot \frac{l^2}{v^2}$$
$$x = \tfrac{1}{2}a't^2 = \tfrac{1}{2}\frac{Bqv}{m} \cdot \frac{l^2}{v^2}$$

FIG. 15.4 Electrons in parallel fields of B and E.

as shown. Further, both are perpendicular to the velocity v and hence do not affect the motion in the direction of v. Thus time t to reach end of plates $= l/v$.

Motion in direction Oy.

$$\text{The displacement } y = \tfrac{1}{2}at^2 = \tfrac{1}{2}\frac{Eq}{m}\cdot\frac{l^2}{v^2} \quad\ldots\ldots (1)$$

Motion in direction Ox.

$$\text{The displacement } x = \tfrac{1}{2}a't^2 = \tfrac{1}{2}\frac{Bqv}{m}\cdot\frac{l^2}{v^2} = \tfrac{1}{2}\cdot\frac{Bq}{m}\cdot\frac{l^2}{v} \quad\ldots\ldots (2)$$

Note the presence of v^2 in (1) and v in (2). Eliminating v, then

$$x^2 = \frac{q}{m}\cdot\frac{B^2 l^2}{2E}\cdot y$$

This is the equation of a parabola. All particles with the same *charge/mass* ratio, q/m, lie on the same parabola, although their velocities may be different (p. 131).

EXAMPLES

1. Electrons in a cathode-ray tube are accelerated through a p.d. of 2 kV between the cathode and screen. Find the velocity with which they strike the screen. If 10^{12} electrons per second strike the screen, and all lose their energy, find (*i*) the average force on the screen, (*ii*) the power dissipated.

With the usual notation,

$$\text{energy of electron } = \tfrac{1}{2}mv^2 = eV$$

$$\therefore\quad v = \sqrt{2\frac{e}{m}V}$$

From $e = 1{\cdot}6\times 10^{-19}$ C, $m = 9{\cdot}1\times 10^{-31}$ kg,

$$v = \sqrt{2\times\frac{1{\cdot}6\times 10^{-19}}{9{\cdot}1\times 10^{-31}}\times 2000}$$

$$= 2{\cdot}7\times 10^7 \text{ m s}^{-1} \quad\ldots\ldots (i)$$

(*i*) $\qquad$ Force = change in momentum (mv) per second

$$= 10^{12}\times 9{\cdot}1\times 10^{-31}\times 2{\cdot}7\times 10^7$$

$$= 24\times 10^{-12} \text{ N}$$

(*ii*) Power dissipation = energy dissipated per second

$$= Q \text{ per second} \times V$$

$$= 10^{12} \times 1 \cdot 6 \times 10^{-19} \times 2000$$

$$= 3 \cdot 2 \times 10^{-4} \text{ W.}$$

2. An oil drop, mass $3 \cdot 25 \times 10^{-12}$ g, falls vertically with uniform velocity between two parallel vertical plates 2 cm apart. When a p.d. of 1000 V is applied to the plates, the path of the drop is then inclined at 45° to the vertical. Calculate the charge on the drop. Calculate the new charge on the drop when its path changes to 26·5° to the vertical.

The force on the drop due to an applied p.d. between the plates is horizontal. Hence the vertical component of its velocity is unaltered.

With a p.d. of 1000 V, the potential gradient $= 1000$ V$/2 \times 10^{-2}$m $=$ intensity E.

$$\therefore \quad \text{force on charge } q \text{ on drop} = \frac{1000}{2 \times 10^{-2}} \times q \text{ newton}$$

Since the path is 45° to vertical,

$$\text{force} = \text{weight of drop}$$

$$= 3 \cdot 25 \times 10^{-15} \text{ kgf}$$

$$= 3 \cdot 25 \times 10^{-15} \times 9 \cdot 8 \text{ newton } (1 \text{ kgf} = 9 \cdot 8 \text{ N})$$

$$\therefore \quad \frac{1000 \times q}{2 \times 10^{-2}} = 3 \cdot 25 \times 10^{-15} \times 9 \cdot 8$$

$$\therefore \quad q = \frac{3 \cdot 25 \times 10^{-15} \times 9 \cdot 8 \times 2 \times 10^{-2}}{1000}$$

$$= 6 \cdot 4 \times 10^{-19} \ (= 4e)$$

(*iii*) When the angle is 26°30′ to the vertical, the charge has decreased to q_1 say. Then, if E is the intensity between the vertical plates,

$$\tan 26°30′ = \frac{Eq_1}{mg} = \frac{Eq_1}{Eq} = \frac{q_1}{q} = \frac{q_1}{4e}$$

$$\therefore \quad q_1 = \tan 26°30′ \times 4e = 0 \cdot 5 \times 4e = 2e.$$

Experiments on electrons. Perrin tube.   The sign of the charge on electrons can be found by deflecting electrons into a Faraday cage and using a connected gold-leaf electroscope to determine the sign of their charges. Fig. 15.5. Experiment shows that they carry *negative* charges, as a negatively-charged leaf rises.

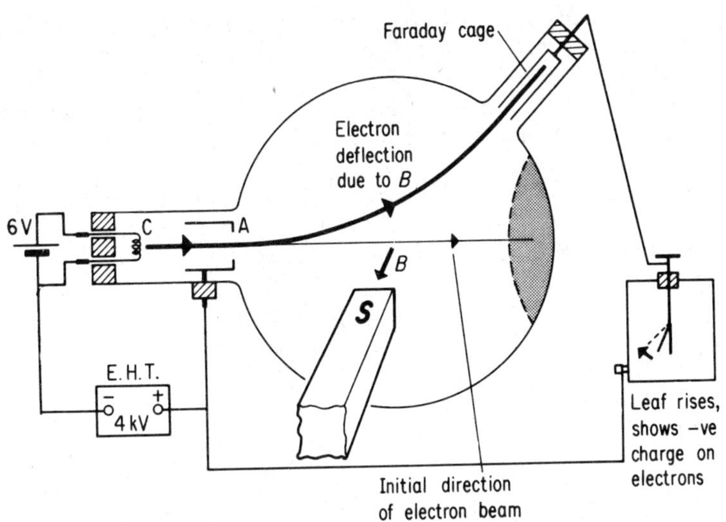

FIG. 15.5 Sign of charge on electron.

Thomson's experiment for e/m of electron. *Apparatus.* Discharge tube at low pressure; perforated anode to allow a fine beam YX of cathode rays to pass between horizontal plates providing a perpendicular electrical field E; two Helmholtz coils S, R to provide a perpendicular magnetic field B by current in them. Fig. 15.6.

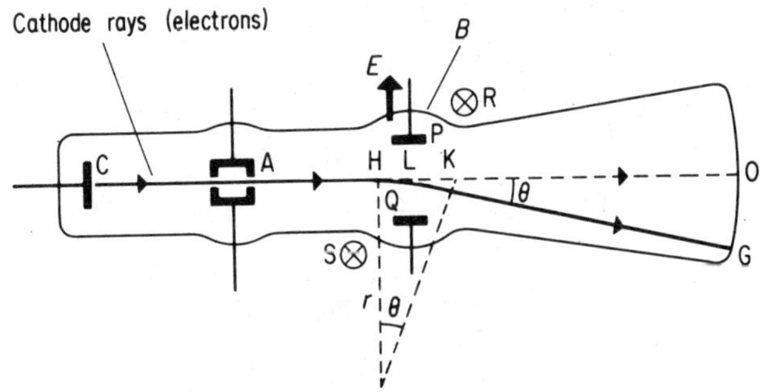

(i) Balanced forces: $Bev = Ee$, $v = \dfrac{E}{B}$

(ii) B alone: $Bev = \dfrac{mv^2}{r}$

$$\therefore \quad \frac{e}{m} = \frac{v}{rB}$$

$$\left(r: \ \tan\theta = \frac{HK}{r} = \frac{OG}{LO} \right)$$

FIG. 15.6 Thomson's experiment for e/m of electron.

127

Measurements. (1) Apply B so that the beam is deflected. Apply E to neutralize the deflection. Then $Ee = Bev$, or $v = E/B$.

Units: v in metre second^{-1} when E in V m^{-1}, B in T (or Wb m^{-2}) Results showed v was of the order 10^6 m s^{-1}. Cathode rays could not be an electromagnetic wave since this has a velocity of about 3×10^8 m s^{-1}.

(2) With B alone, the deflection OG or x was measured. Then $Bev = mv^2/r$, or $e/m = v/rB$. Now, from Fig. 15.6, $\tan \theta = $ HK$/r$ $= x/$LO or $r = $ HK.LO$/x$. Hence r was found. Since $e/m = v/rB$, then e/m was calculated.

Units: e/m in coulomb kilogramme^{-1} (C kg^{-1}) when v in m s^{-1} r in m, B in T.

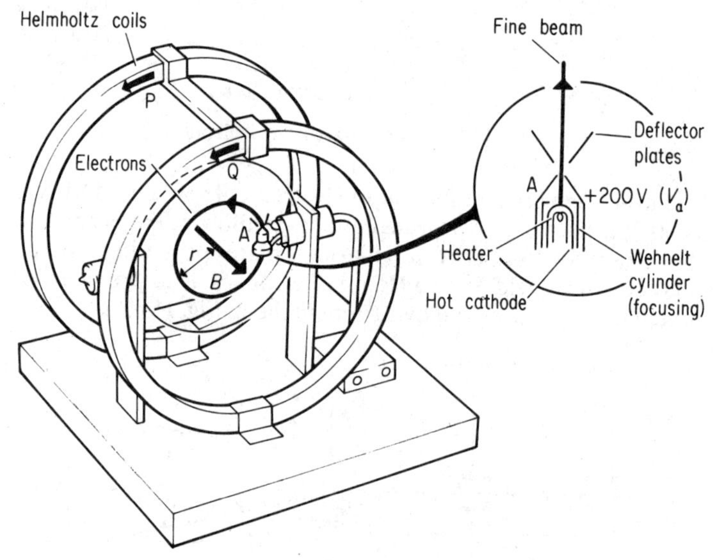

$$\text{Deflection:} \quad Bev = \frac{mv^2}{r}$$

$$\therefore \frac{e}{m} = \frac{v}{rB} \quad \cdots \quad (1)$$

$$\text{Velocity:} \quad \tfrac{1}{2}mv^2 = eV_a \quad \cdots \quad (2)$$

$$\text{Eliminate } v: \quad \frac{e}{m} = \frac{2V_a}{r^2 B^2}$$

FIG. 15.7 Fine beam tube method for e/m.

Deduction of electron mass. For an electron, $e/m = 1\cdot 8 \times 10^{11}\,\mathrm{C\,kg^{-1}}$. The e.c.e. of hydrogen is about $1 \times 10^{-8}\,\mathrm{kg\,C^{-1}}$, so that for a hydrogen ion $m_H/e = 1 \times 10^{-8}\,\mathrm{kg\,C^{-1}}$ *assuming* the charge on a hydrogen ion is e.

$$\therefore \quad \frac{m_H}{e} \times \frac{e}{m} = 1 \times 10^{-8} \times 1\cdot 8 \times 10^{11} = 1800 = \frac{m_H}{m}$$

Thus the hydrogen ion is about 1800 times as heavy as the electron.

Fine beam tube method for e/m. In this modern method, (*i*) electrons are liberated by a hot cathode, (*ii*) Helmholtz coils P, Q provide a uniform known magnetic field B to produce a circular deflection, radius r. Fig. 15.7. The anode V_a enables the velocity v to be found, since $\frac{1}{2}mv^2 = eV_a$, and hence e/m is obtained.

Millikan's experiment for e.   *Apparatus.* Two parallel plates about a centimetre apart, the upper one with a fine hole in the centre; an oil spray above the hole; a constant temperature enclosure; a battery across the two plates which can be switched on or off or reversed; a powerful small source of light to illuminate the space between the plates; an X-ray machine which can be switched on to ionise the air between the plates; a low-power telescope with a calibrated eyepiece, to observe through a window light scattered from an oil drop between the plates. Fig. 15.8.

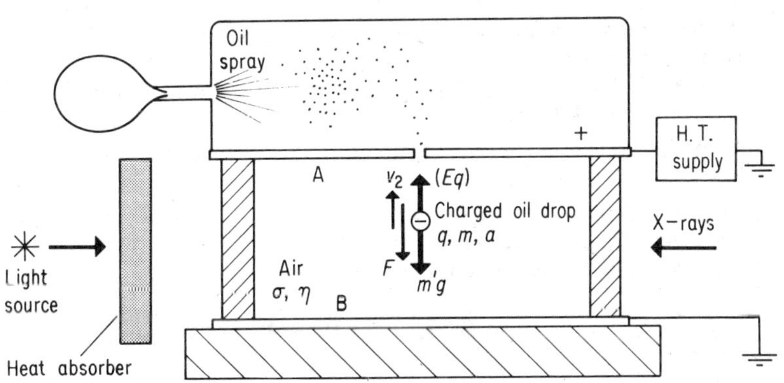

Without field: $m'g = \frac{4}{3}\pi a^3 (\rho - \sigma) g = 6\pi \eta a v_1 \cdots (1)$

With field : $Eq - m'g = 6\pi \eta a v_2 \cdots (2)$

$$\therefore \quad q = \frac{6\pi \eta a}{E}(v_1 + v_2)$$

Result : e = lowest common factor of $q = 1\cdot 6 \times 10^{-19}\,\mathrm{C}$

FIG. 15.8 Millikan's experiment for e

Method. A drop falls through the hole and is timed as it falls under gravity. The electric field is switched on to keep the same drop in view. The drop is again timed as it rises against gravity with the aid of an applied electric field. The same drop is timed again after collecting more charges.

Theory. (*i*) Under gravity; suppose the velocity is v_1. Then, using Stokes' law,

$$\tfrac{4}{3}\pi a^3(\rho - \sigma)g = 6\pi\eta a v_1.$$

(*ii*) Applied electric field E; suppose the velocity is v_2 upwards. Then if q is the charge on the drop

$$Eq - \tfrac{4}{3}\pi a^3(\rho - \sigma)g = 6\pi\eta a v_2.$$

$$\therefore \quad Eq = 6\pi\eta a(v_2 + v_1)$$

$$\therefore \quad q = \frac{9\pi\eta^{3/2}}{E}\left[\frac{2}{g(\rho-\sigma)}\right]^{\frac{1}{2}} v_1^{1/2}(v_1 + v_2).$$

Conclusion. Results show that q is an *integral* multiple of a basic charge e equal to about $1{\cdot}6 \times 10^{-19}$ C. This charge e, the smallest possible charge, is the charge on an electron.

Errors. (1) The viscosity of air at the particular temperature was not known accurately. (2) Stokes' law required correction—this law applies to a continuous medium, but the tiny oil-drops fall through "holes" in the air as their size is comparable with the mean free path of the molecules.

e from electrolysis.   If N_A is the Avogadro constant, then, for a monovalent element such as silver, number of silver ions in solution in silver nitrate solution $= N_A$. The Faraday constant, F, is the charge required to deposit N_A ions $=$ charge carried by N_A ions. Thus if a monovalent element carries a charge e, then $N_A e = F$.

$$\therefore \quad e = \frac{F}{N_A} = \frac{96\,500\text{C}}{6{\cdot}02 \times 10^{23}} = 1{\cdot}6 \times 10^{-19}\text{C (approx.)}$$

Mass spectrometers. (1) *Thomson "positive rays" apparatus.* Positive rays (ions) produced in a discharge tube at low pressure. They passed through a very fine tube C, the cathode Fig. 15.9a. The ions were then deflected by *parallel* electric (E) and magnetic (B) fields produced between the insulated poles of a magnet, both perpendicular to the direction of the ions. The ions were incident on a photographic plate P. Several parabolas were produced. Fig. 15.9b.

Theory. Suppose E and B are uniform over a length l, and q, M are the charge and mass of a positive ion moving with velocity v on entering the fields.

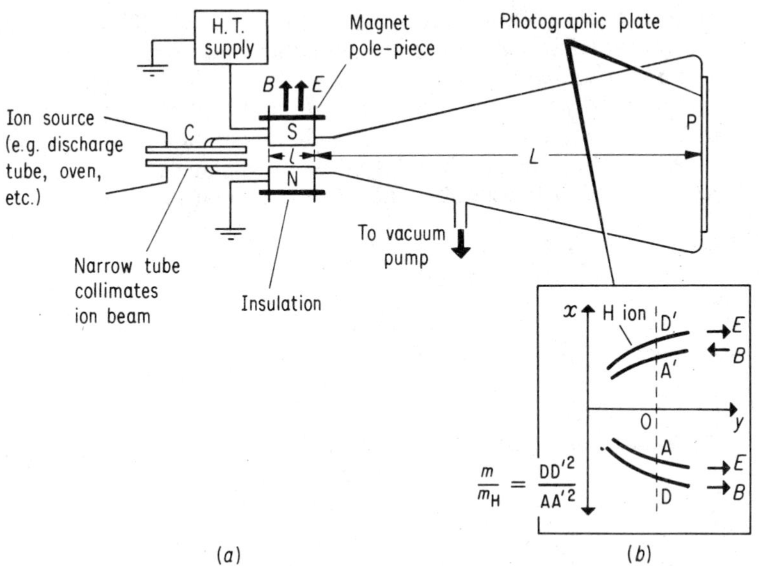

FIG. 15.9 Thomson's mass spectrometer.

Electric field effect: Acceleration of ion due to $E = Eq/M$. Time to travel in field $= l/v$. Hence vertically velocity reached $= (Eq/M) \times l/v = Eql/Mv$. On leaving the field the ion travels in a straight line to the plate P, a distance L away. Time taken to travel $L = L/v$.

$\therefore$ vertical displacement $y = (Eql/Mv) \times (L/v) = EqlL/Mv^2$ (i)

Magnetic field effect: Acceleration of ion due to $B = Bqv/M$. Time to travel in field $= l/v$. Hence horizontal velocity reached, perpendicular to B, $= Bqvl/Mv$. Time to travel $L = L/v$.

$\therefore$ horizontal displacement $x = BqlL/Mv$ . . (ii)

Eliminate v between (i), (ii). Then

$$x^2 = \frac{q}{M}\left(\frac{B^2lL}{E}\right)y \quad . \quad . \quad . \quad . \quad . \quad . \quad (iii)$$

This is the equation of a parabola. Ignoring the small displacement occurring in the region of two fields, *all* ions with the same ratio q/M will lie as the same parabola, although each may move with different velocity.

Determination of masses of ions. The lightest particle such as a hydrogen ion corresponds to the outermost parabola. Since $q/M \propto x^2$ for a given value of y, $m/m_H = x_H^2/x^2$, where m, m_H are the masses of an unknown ion and a hydrogen ion respectively and x, x_H are the respective displacements. Thus m can be found. Thomson found two parabolas for chlorine, for example. Thus chlorine has atoms of different masses although they have the same chemical properties (*isotopes*, p. 149).

(2) *Bainbridge.* Much more accurate and sensitive than Thomson's. Positive ions, produced in a discharge tube, pass through slits S_1, S_2 (Fig. 15.10). They passed between plates P, Q, having an electric field E between them and a perpendicular magnetic field B_1.

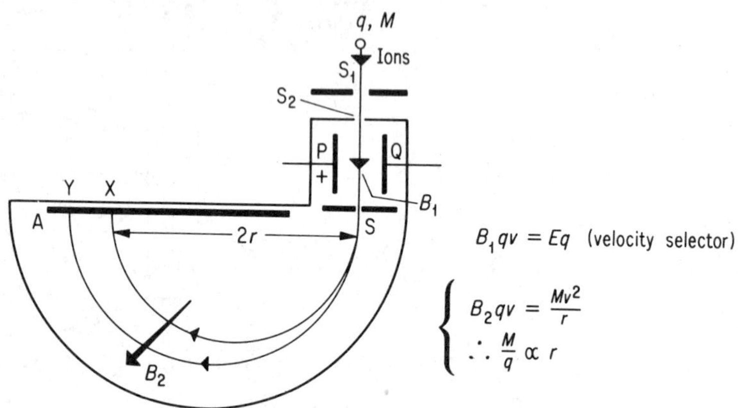

FIG. 15.10 Bainbridge mass spectrometer.

Ions of charge q which have a velocity v given by $B_1qv = Eq$, or $v = E/B_1$, can pass undeflected through P, Q and through a slit S. The arrangement is therefore a *velocity selector*. The selected ions are now deflected by a magnetic field B_2, and travel in a circular path and are photographed on a plate A.

Theory.
$$B_2qv = Mv^2/r,$$

or
$$\frac{M}{q} = \frac{B_2r}{v} = \frac{B_1B_2r}{E}.$$

Hence for ions with the same charge q, $M \propto r \propto 2r$. Thus $\triangle M \propto \triangle (2r)$. The change in $2r$, or $SY - SX$, for ions of different mass is therefore directly proportional to the mass change. The Bainbridge spectrometer has the considerable advantage of a linear mass scale.

16. RADIO VALVE. JUNCTION DIODE. TRANSISTOR

A *diode* is a "non-ohmic" device; it has low resistance in one direction and a high resistance in the opposite direction. It is therefore used (*i*) as a "rectifier"—it converts a.c. to d.c. voltage, (*ii*) as a "detector"— it enables the a.f. energy carried by modulated radio waves to be detected.

Diode valve. Uses electrons, liberated by a hot cathode C inside a glass envelope containing a vacuum. Fig. 16.1*a*. The electrons move towards a nickel anode plate A when this has a positive potential relative to C and then flow in the outside circuit. The valve does not conduct when A is negative in potential relative to C.

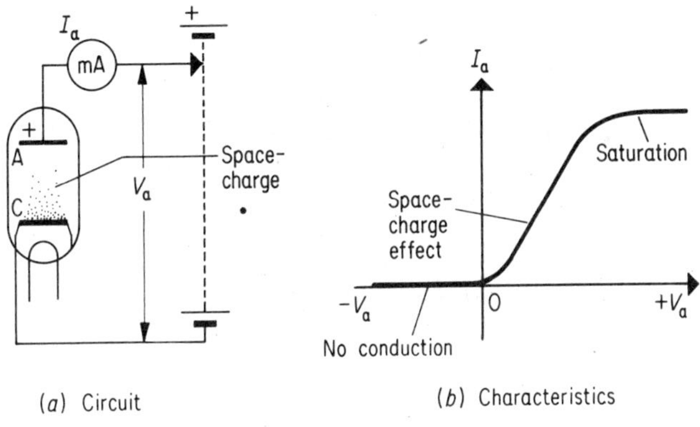

(*a*) Circuit   (*b*) Characteristics

FIG. 16.1 Diode valve. Characteristic.

Characteristic. The $I_a - V_a$ curve has a small initial curvature followed by a straight line; the curve flattens to a horizontal line, i.e. saturation occurs, when the anode voltage is high enough. Fig. 16.1*b*.

Explantion. Before saturation occurs, all the electrons emitted by the cathode do not reach the anode. This is due to the repulsive effect of the negative *space charge*, the electrons moving between the cathode and anode. When the anode potential V_a is sufficiently large, the effect of the space charge is completely overcome. The number of electrons reaching the anode is then equal to the number emitted by the cathode, which depends on the cathode temperature.

Junction diode. A semiconductor solid device. Uses a "p-n junction" formed at the boundary between a p-semiconductor (one with majority positive carriers called "holes", produced by doping a pure semiconductor such as silicon or germanium with indium) and a n-semiconductor (one with majority negative carriers, electrons, produced by doping a pure semiconductor with antimony). Fig. 16.2*a*.

The p-n junction conducts well when the p-semiconductor is more than about 0·5 V positive relative to the n-semiconductor. It has a very much higher resistance when the p.d. is reversed. Fig. 16.2*b*.

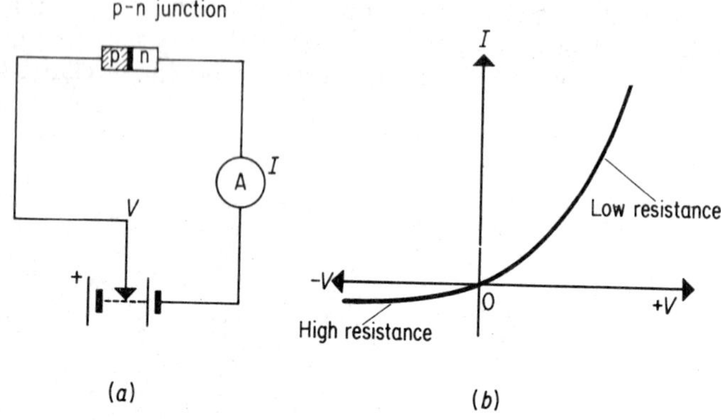

Forward bias (*V* +ve): good conduction, due to majority carriers
Reverse bias (*V* −ve): poor conduction, due to minority carriers

FIG. 16.2 Junction diode.

Rectification. Fig. 16.3*a* illustrates the principle of converting a.c. voltage (V_{in}) to d.c. voltage (V_{out}). This is *half-wave* rectification.

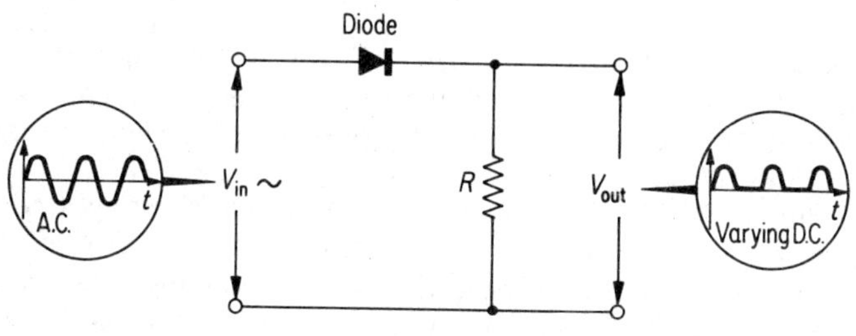

Diode conducts well on one half cycle only

(a) Half—wave

FIG. 16.3. Rectification of a.c. voltage.

To provide a stable d.c. voltage, *full-wave* rectification by two diodes are needed—each conducts in turn—followed by a *filter circuit* which removes the "ripple" from the varying d.c. voltage. Fig. 16.3*b*.

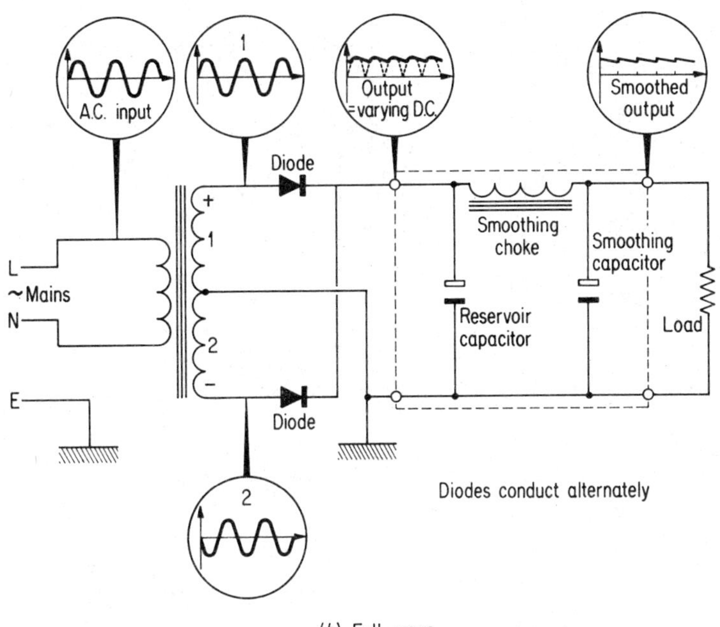

(b) Full-wave

FIG. 16.3 Rectification of a.c. voltage.

TRIODE VALVE

This three-electrode thermionic device has (*i*) a hot cathode C liberating electrons, (*ii*) an anode A, (*iii*) a wire grid G between A and C. See Fig. 16.4*a*.

When the grid potential varies it has a sensitive control over the electron flow towards the anode. An alternating voltage of 0·1 volt (r.m.s.) applied between the grid and cathode may therefore produce an equivalent effect of 2 volts (r.m.s.) in the anode circuit under suitable conditions. The voltage gain is then 20, and the triode has acted as an *amplifier*.

Characteristics. Fig. 16.4*a* shows a circuit for investigating how anode current I_a varies when V_g is changed (V_a constant) and when V_a is varied (V_g constant). Fig. 18.6*b* illustrates typical results.

Valve constants (parameters) (1) A.C. or slope resistance, R_a, $= \delta V_a/\delta I_a$; obtained from slope of anode characteristic, $I_a - V_a$ curve, V_g constant. Thus, from Fig. 16.4*b*, for $V_g = 0$, $\delta I_a = 8 - 4 = 4$ mA when $\delta V_a = 150 - 110 = 40$V. Hence $R_a = 40$V/4 mA $= 10\,000\,\Omega$.

(2) Mutual conductance, $g_m = \delta I_a/\delta V_g$, V_a constant; obtained from slope of mutual characteristic, $I_a - V_g$ curve, V_a constant. Thus from

135

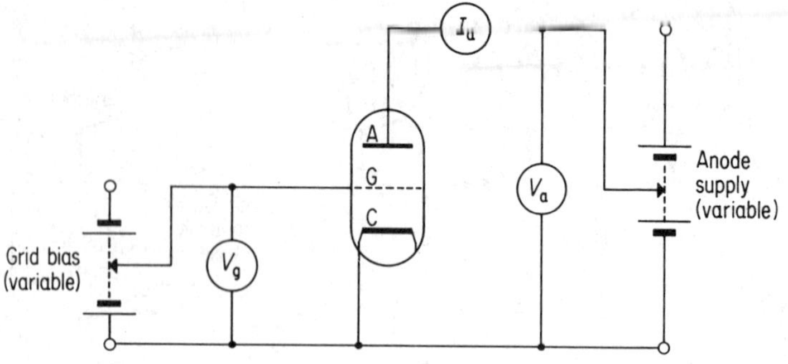

(a) Circuit for characteristics

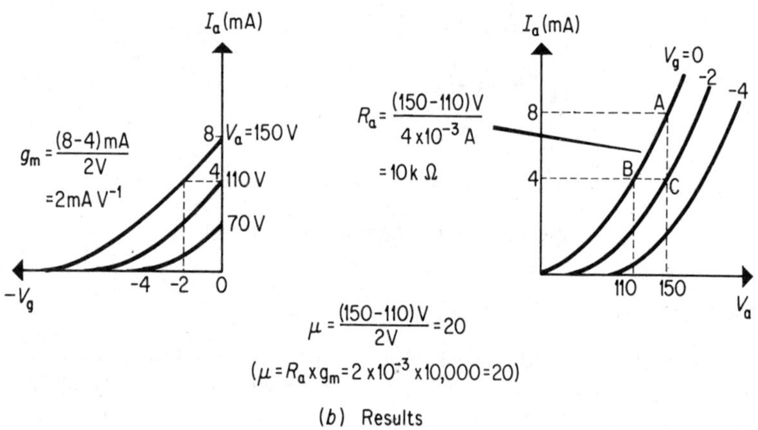

(b) Results

Fig. 16.4 Triode valve characteristics and constants.

Fig. 16.4b, when $V_a = 150$ V, we have $\delta I_a = 8 - 4 = 4$ mA when $\delta V_g = 2$ V. Hence $g_m = 4$ mA/2 V $= 2$ mA V^{-1}.

(3) Amplification factor, $\mu = \delta V_a/\delta V_g$, δI_a constant; obtained either from anode or mutual characteristics. Thus from Fig. 16.4b, using ABC, a change of 4 mA is obtained for $\delta V_a = 150 - 110 = 40$ V ($V_g = 0$), and for $\delta V_g = 0 - (-2) = 2$ V($V_a = 150$ V). Hence $\mu = 40$ V/2 V $= 20$. Note that

$$\mu = R_a \times g_m,$$

when R_a in Ω and g_m in A V^{-1}. Thus, from above, since $R_a = 10\,000\,\Omega$, $g_m = 2 \times 10^{-3}$ A V^{-1}, we have $\mu = 10\,000 \times 2 \times 10^{-3} = 20$.

TRANSISTOR

A transistor is a three-terminal solid state device. A *p-n-p* transistor has a p-emitter, E; a n-base, B; a p-collector, C. Fig. 16.5*a*. The base is very thin; the collector area is large compared to the emitter area.

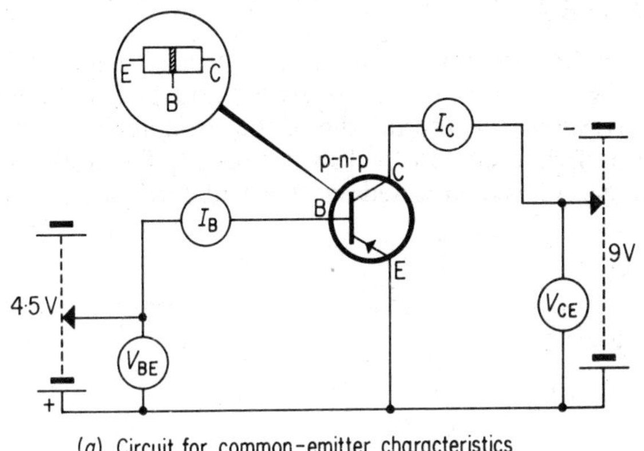

(*a*) Circuit for common-emitter characteristics

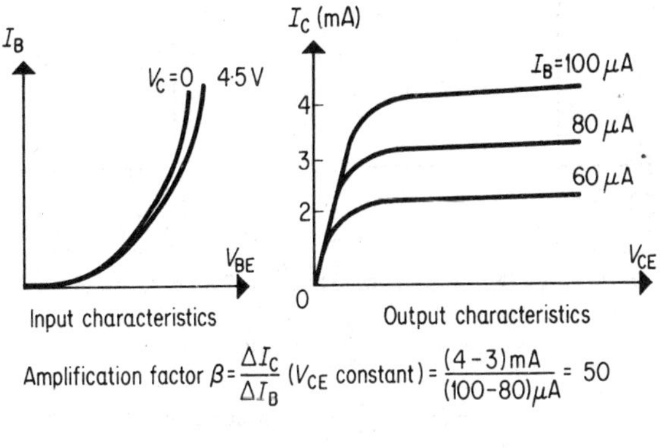

Amplification factor $\beta = \dfrac{\Delta I_C}{\Delta I_B}$ (V_{CE} constant) $= \dfrac{(4-3)\,\text{mA}}{(100-80)\,\mu\text{A}} = 50$

(*b*) Results

FIG. 16.5 Transistor and characteristics.

Circuit: The emitter-base (E−B) circuit is always "forward biased" so that a large number (majority) of carriers are injected towards the base and collector. Thus in a *p-n-p* transistor, E has a positive potential relative to B (compare the *p-n* junction diode, p. 134). The collector-base (C−B) base is always *reverse*-biased. The grounded- or *common-emitter* circuit is most widely used as an amplifier circuit; this has the emitter common to the input (base-emitter) and output (collector-emitter) circuits. Fig. 16.5*a*.

Characteristics: A circuit for investigating common-emitter (CE) characteristics is shown in Fig. 18.5a. Typical curves are shown in Fig. 16.5b.

17. AMPLIFIERS. OSCILLATORS. OSCILLOGRAPH

Triode as voltage amplifier. Circuit: (1) A load such as a resistance R is used in the anode circuit. (2) The steady grid-bias (G.B.) is negative, and corresponds to the middle of the straight-line part of the $I_a - V_g$ (mutual) characteristic—class A bias. (3) The applied alternating voltage, V_i, is connected in the grid-cathode circuit (Fig. 17.1a).

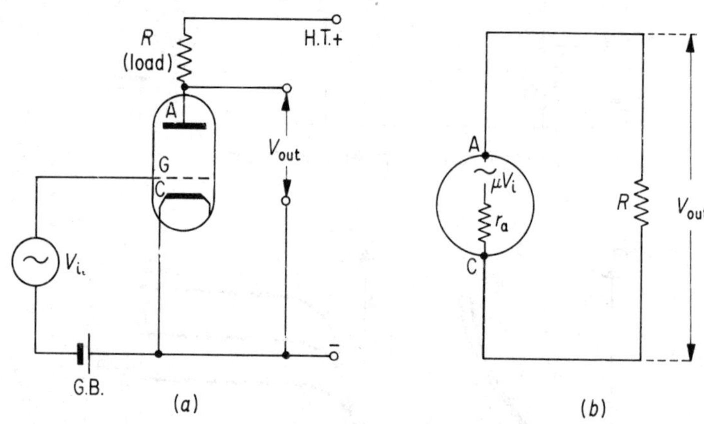

With class A G.B., valve acts as A.C. generator of e.m.f. μV_i

$$V_{out} = \frac{R}{R + r_a} \times \mu V_i$$

$$\therefore \frac{V_{out}}{V_i} = \frac{\mu R}{R + r_a} = \text{voltage gain}$$

FIG. 17.1 Triode voltage amplifier.

Action: If the grid potential varies along the linear part of the characteristic and is always negative, the output *alternating* voltage V_0 across R is undistorted (class A amplification). The valve then acts as an alternating voltage generator of e.m.f. μV_i and internal resistance r_a Fig. 17.1b. As this is connected to the resistance R, then $V_0 = IR = \mu V_i R/(R+r_a)$.

$$\therefore \quad \text{voltage gain or amplification factor} = \frac{V_0}{V_i} = \frac{\mu R}{R+r_a}.$$

Thus if $\mu = 20$, $r_a = 10\ 000\ \Omega$, $R = 15000\ \Omega$, then voltage gain $= 20 \times 15\ 000/25\ 000 = 12$.

Triode as Oscillator. The essential features are: (1) The oscillatory circuit (coil-capacitor), $L, R - C$, in the anode circuit. (2) The tapping on the coil in the anode circuit feeds back energy for the oscillatory circuit—the grid-cathode voltage is amplified and fed back in phase with the anode current. (3) An automatic grid-bias, $C_g - R_g$, providing a large negative grid-bias. (4) A d.c. supply such as an H.T. battery (Fig. 17.2).

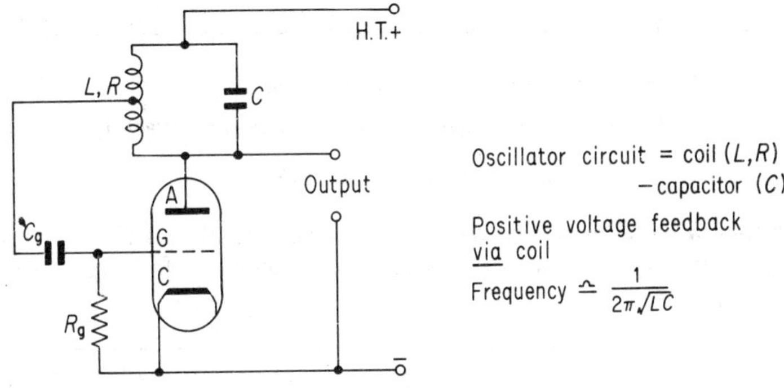

FIG. 17.2 Triode oscillator.

When the circuit is made, oscillations of frequency about $1/2\pi\sqrt{LC}$ occur in the coil-capacitor circuit. Feedback compensates for the energy dissipated in the resistance R, which would damp the oscillations. With correct bias conditions, the efficiency of the circuit (a.c. power in the oscillator circuit $\div$ d.c. power from the battery) is high.

Transistor amplifier. The transistor is a current-amplifier, whereas the triode valve is a voltage amplifier. A typical common-emitter a.f. amplifier circuit is shown in Fig. 17.3. R_1 and R_2 form a potential divider across the supply battery V_{CC} and maintain the necessary base-emitter and collector-emitter bias. R is the load resistance in the collector circuit (compare the load resistance for the triode valve). R_3 is the emitter resistance, which prevents excessive temperature rise—an increase in collector current increases the p.d. across R_3 and decreases the base-emitter bias, thus counteracting the rise in collector current. C_2 is a large by-pass capacitor—it by-passes the a.c. component of the output, preventing feedback to the input circuit which would cause distortion. C_1 allows only a.c. from the input to reach the base circuit. C_3 allows only a.c. to reach the output circuit.

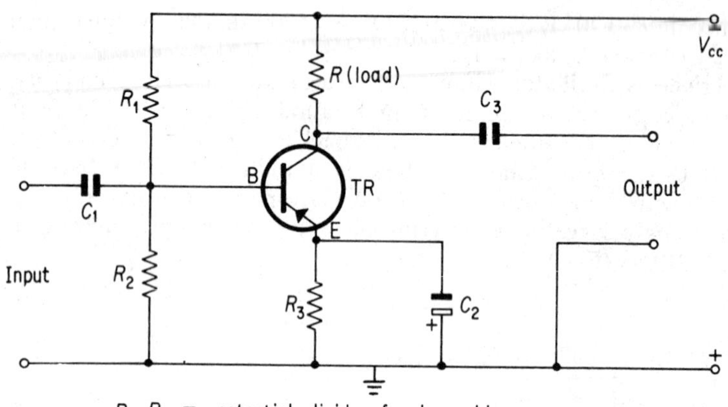

R_1, R_2 = potential divider for base bias

R_3 = emitter resistance, limits excessive current I_c

FIG. 17.3 Transistor amplifier.

Transistor oscillator. The oscillatory circuit, L, R and C, is in the collector circuit. L feeds back energy to compensate for that dissi-

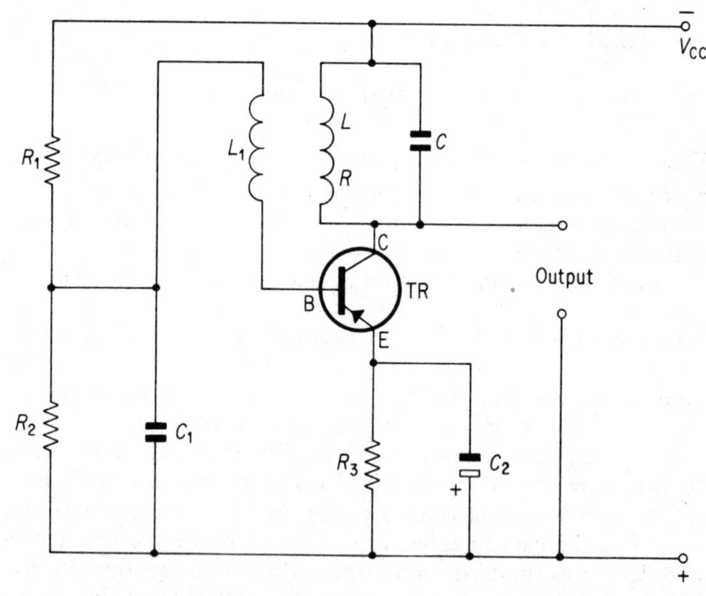

Oscillator circuit = coil (L,R) − capacitor (C)
R_1, R_2 = potential divider for base bias
Positive current feedback via L_1

FIG. 17.4 Transistor oscillator.

140

pated in this circuit. Fig. 17.4. Other components have a similar purpose to those in the amplifier circuit in Fig. 17.3. Compare also the *valve oscillator*, Fig. 17.2.

CATHODE-RAY OSCILLOGRAPH

Contents. A *cathode*, emitting electrons; a metal cylinder or plate with a hole in the middle, the *grid*, G; two or three cylinders, A_1, A_2, A_3, the *anodes*; a pair of *X-plates*; a pair of *Y-plates*; a *screen* coated with zinc sulphide; the whole contained in an evacuated glass tube.

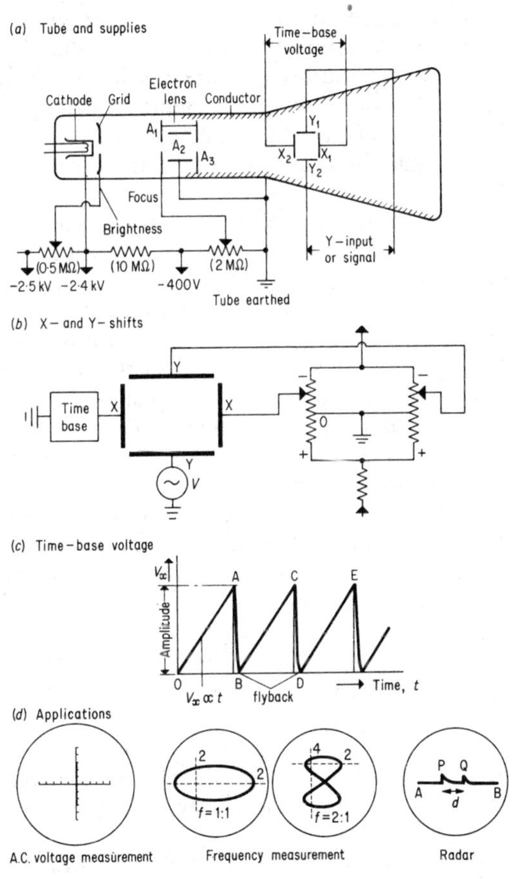

FIG. 17.5 Cathode ray oscillograph.

Voltage supplies and relative potentials. Fig. 17.5a. If the cathode has zero potential, then the grid is several volts negative and the anodes are a few thousand volts positive, The inside of the glass tube is coated with a graphite paint and joined to an anode, so that the screen is at a high voltage. In practice, the tube and the anode are earthed for safety; the cathode is thus a few thousand volts negative relative to the tube.

Controls. The *brilliance* is controlled by the grid potential; the more negative it is relative to C, the fewer the electrons passing it. The *focus* is controlled by the potential of the anode A_1, which alters the focal length of the electron lens comprising A_1, A_2, A_3. The X-*shift* and Y-*shift* shift the beam in a horizontal and vertical direction respectively, by making the potentials of the respective X- and Y-plates more negative or positive as the case may be. Fig. 17.5*b*. The voltage V_y to be examined is connected to the Y-plates. The time-base is already in the oscillograph, and is connected to the X-plates when a stationary picture is required. By using a *trigger* control, the frequency of the time-base voltage V_x can be synchronised with the Y-signal, so that a stationary trace is obtained on the screen. Fig. 17.5*c*.

Uses of oscillograph. (1) D.c. and a.c. voltages can be measured by first measuring the length or height of the trace on the screen when a known voltage is applied to the Y-plates, so that the voltage per millimetre length of trace is obtained. Fig. 17.5*d*. (2) Frequency can be measured by connecting a signal generator S to the X-plates and the unknown frequency variation T to the Y-plates. The frequency of S is varied until an ellipse or circle is obtained. The frequency of T is then equal to the frequency of S, read from its scale. (3) Phase difference can be measured by means of a Lissajous' figure. (4) In Radar, small time intervals between two signals P, Q can be measured by receiving them on the screen, and measuring their distance d apart. This can be compared with the length of the whole trace AB, and the time-interval is then deducted from the known frequency of the time-base. Fig. 17.5*d*.

18. RADIOACTIVITY

(1) Radioactivity is a property of the *atom* of an element due to the instability of its nucleus. Radioactive substances emit α- and β-particles and γ-rays.

Some detectors of ionizing particles. (1) *Geiger-Muller (GM) tube* (Fig. 18.1). In one form it consists of a tube with a thin window M, containing a tungsten wire. The walls are coated with graphite to make them conducting. The tube is filled with a mixture of argon and bromine vapour at low pressure. The wire is kept at a high positive potential V, about 400V, corresponding to the plateau of its charac-

	α-PARTICLE	β-PARTICLE	γ-RAY
Nature	Helium nucleus. ($_2^4$He)	Electron. (e^-)	Electromagnetic wave of very short wavelength, less than 10^{-10} m.
Charge	Positive ($+2e$).	Negative ($-e$).	—
Ionizing power in air	Strong.	Small.	Weak.
Penetrating power	Low—easily absorbed by thin foil or paper.	High.	Very high—thick lead blocks to absorb rays.
Range in air at normal pressure	Few cm.	Order of a metre.	Greater than β-particle.
Other points	Produces scintillations on ZnS screen. Weak effect on photographic plate. Detected by solid state detector, cloud chamber, scintillations, Geiger-Müller tube with thin window, pulse electroscope. Used in early experiments for nuclear disintegration. Deflected slightly in magnetic field, showing heavy particle with +ve charge.	High velocities. Tracks in air photographed in cloud chamber. Detected by Geiger-Müller tube, pulse electroscope. Deflected considerably in magnetic field, showing very light particle with −ve charge.	Detected by Geiger counter. Unaffected by magnetic field, showing wave property.

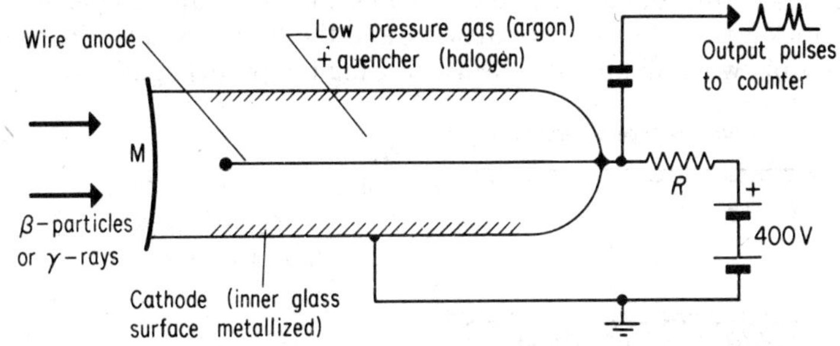

FIG. 18.1 Geiger-Muller tube.

teristic curve, so that the number of pulses produced per second is independent of the voltage. When a single ionizing particle enters the counter, a few electrons and ions are produced in the gas. The potential V is above the breakdown potential of the gas. The ions and electrons are then multiplied rapidly by "ionization by collision". The discharge between the wire and tube produces a current in the large resistance R. The p.d. across R is amplified and passed to a counter, which registers the passage of the ionizing particle. The discharge is *quenched* quickly by the bromine vapour, so that successive particles can be counted. Detects β-particles and γ-rays, and α-particles by using a special thin window.

(2) *Solid state detector.* Energetic α-particles produce more electron-hole pairs near the p-n junction of this semiconductor diode. The increased current is amplified and passed to a counter. Fig. 18.2. Solid state detectors can also detect β-particles and γ-rays.

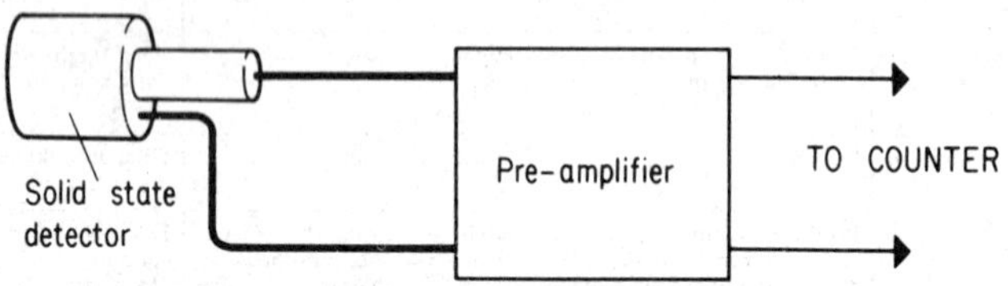

α-particles produce more electron-hole pairs at p-n junction of detector. The current pulse is amplified and passed to a counter

FIG. 18.2 Solid state detector.

(3) *Diffusion cloud chamber.* This contains a region of super-saturated alcohol vapour, produced by cooling the base of the chamber with "dry ice" (solid CO_2). Fig. 18.3. The vapour condenses on ions formed by α- or β-particles, and the trail of minute droplets is visible when illuminated, thus showing the particle path.

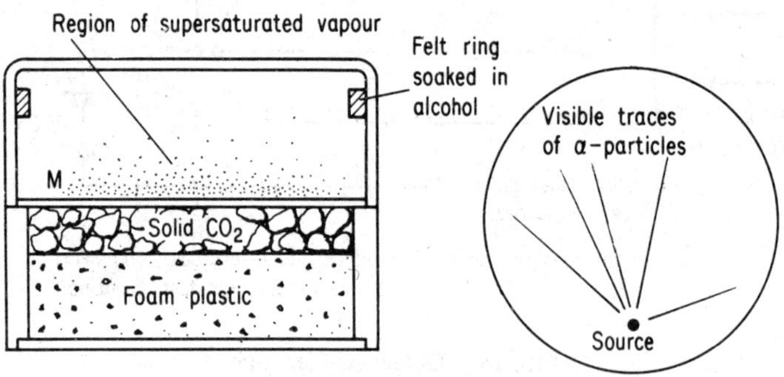

Alcohol vapour sinks. Supersaturated region above cold metal plate M. α-particle tracks shown by condensation on ions

FIG. 18.3 Diffusion cloud chamber.

(4) *Wilson cloud chamber.* Contains supersaturated water-vapour in dust-free air. The air is expanded quickly as it is exposed to ionizing particles, so that cooling occurs. Water droplets condense round ions formed, and are simultaneously illuminated and photographed.

(5) *Bubble chamber* (Glaser). Liquid hydrogen is superheated. When pressure is released, bubbles are not formed for a short period. During this quiet period ionizing particles are introduced. Bubbles are immediately formed round the particles and grow, and are photographed.

(6) *Pulse electroscope.* This consists of a metal "flag" and a neighbouring (side) electrode, each at a high potential. The flag "pulses" when α-particles ionize the air between an external rod or terminal joined to the flag and the outer casing.

(7) *Scintillation photomultiplier.* An ionizing particle strikes a scintillation material or phosphor at one end. The light then falls on a photosensitive material, which emits electrons. These are accelerated towards a metal at $+100V$ say, which emits several times as many electrons. The effect is multiplied rapidly at other metals. The pulse of current, and voltage produced in a resistor, are then detected by electronic circuits.

(8) *Emulsions.* Thick emulsions, with high concentration of silver bromide in gelatine, are used. Very short tracks of silver granules, of the order of a millimetre, are produced.

Scaler. Ratemeter. A *dekatron counter* has (*i*) dekatron tubes, each containing a glow (discharge) which can move round a circular

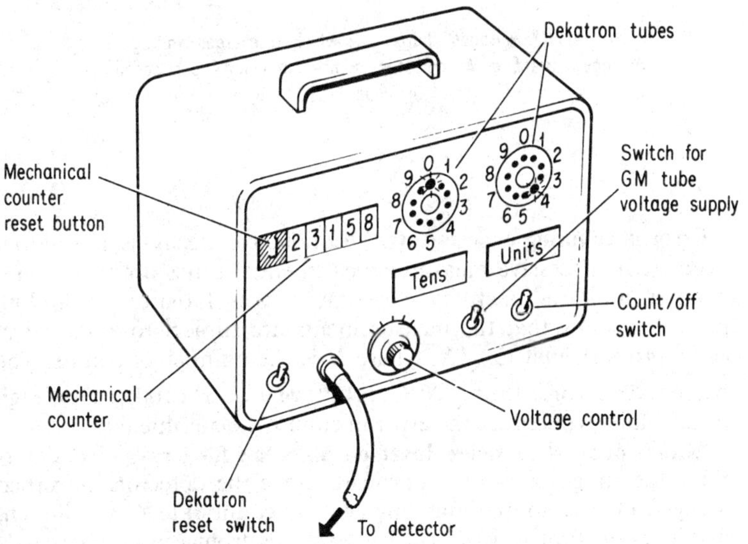

Scalers count number of ionising particles or radiation. Dekatron tubes count to 100 pulses, then pass count to mechanical counter

Fig. 18.4 Scaler.

scale with digits 0–9 (counting units, tens, etc.) and (*ii*) a mechanical counter, which stores counts of 100 or 1000 from the last dekatron tube. Fig. 18.4. The dekatron counter counts individual pulses, i.e. it acts as a *scaler*.

A *ratemeter*, however, provides a reading on a microammeter which is the average number of pulses per second or *count rate*. It has a capacitor C which stores charge when the pulses from a GM tube are passed to it and discharges through a high resistor R in series with the microammeter A. Fig. 18.5. The greater the rate at which pulses are received, the greater is the current, i.e. the count rate $\propto$ current. The time constant CR should be high compared with the interval between successive pulses.

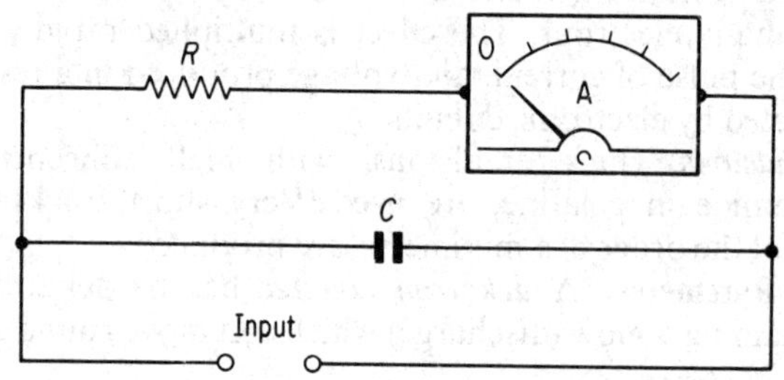

Pulse (input) voltages charge C, which discharges through R.
Average current in A indicates number of counts per second
or count rate

Fig. 18.5 Ratemeter.

Error in counting experiments. Radioactive decay is a random process. If successive measurements in equal times are taken on a counter in a radioactivity experiment, the statistics of random processes show that the measurements are subject to a statistical error proportional to $\sqrt{N}$, where N is the number of counts. The percentage error is thus $\sqrt{N}/N \times 100\%$ or $1/\sqrt{N} \times 100\%$. Thus high total counts will reduce the error in counting experiments.

Range of α-, β-particles. Inverse-square law for γ rays. To determine the range of α- or β-particles, move the detector concerned away from the source and observe the count rate C at different distances, d. Plot C v. d. The curve drops from the flat part to a cut-off value at a distance roughly equal to the range, though this is less precise for β-particles.

The graph of C v. d for γ-rays decreases smoothly. If the rays obey an inverse-square law with distance, then $C \propto 1/(D+a)^2$, where D is the distance from source to detector and a is the unknown constant distance to the actual activated region inside the detector. Fig. 18.6*a*. To see if this is the case, plot $C^{-1/2}$ v. D and see if the points lie on a straight line. Fig. 18.6*b*.

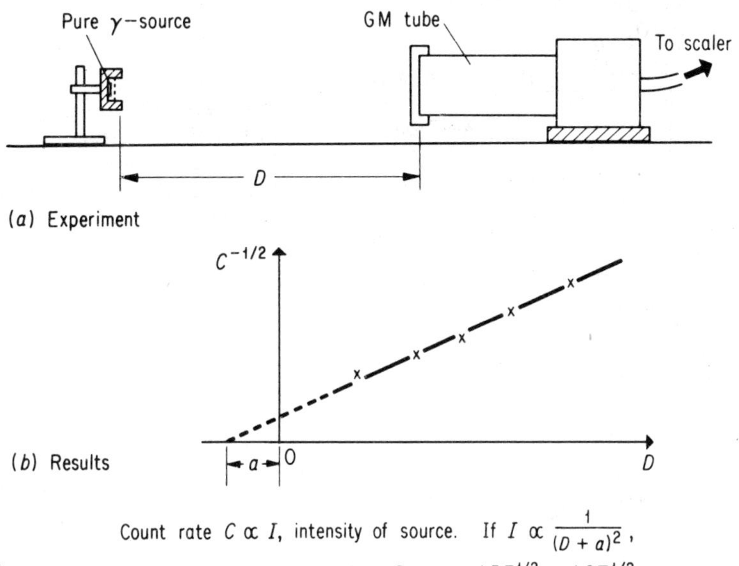

Count rate $C \propto I$, intensity of source. If $I \propto \dfrac{1}{(D + a)^2}$,

where a is a constant, then $D + a = kI^{-1/2} = kC^{-1/2}$

FIG. 18.6 Inverse-square law verification.

Half-life. Decay constant. The atoms of radioactive elements decay according to a statistical law or law of chance. Suppose N is the number of atoms left at an instant. Then, in a time dt, the number of atoms, dN, decaying is given by

$$dN = -\lambda N \,.\, dt,$$

where λ is a constant known as the *decay constant*. Thus $dN/dt = -\lambda N$, and hence

$$N = N_0 e^{-\lambda t}, \quad \ldots \ldots \ldots \ldots \quad (i)$$

where N_0 is the original number of atoms, at $t = 0$.

The *half-life*, $T_{1/2}$, is the time taken for half the atoms to disintegrate. When $N = N_0/2$, then, from (i), $T_{1/2} = \log_e 2/\lambda = 0\cdot693/\lambda$. The half-life, or decay constant, is a property of the particular radioactive atom.

Determination of half-life. Measure the count rate, C, after equal intervals of time t using a ratemeter or scaler. Fig. 18.7*a*. Plot (i)

147

C v. t and find the average time for the count-rate to decrease to one-half, Fig. 18.7b, or (*ii*) log N v. t, draw the best straight line through the points, and find the gradient g of the line. Since log N = log $N_0 - \lambda t$, then $g = -\lambda$; and $T_{1/2} = 0{\cdot}693/\lambda$.

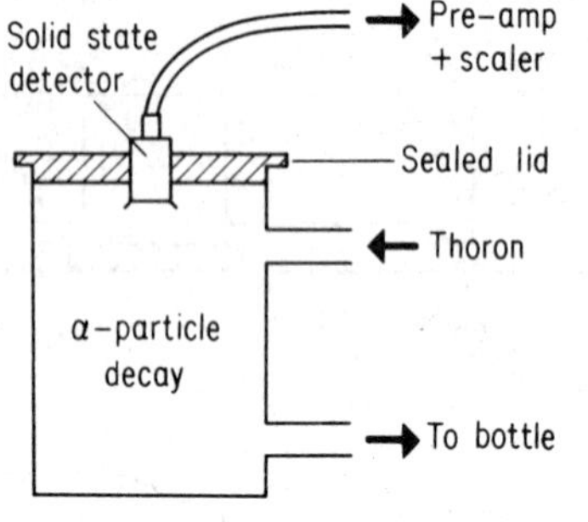

(*a*) Experiment

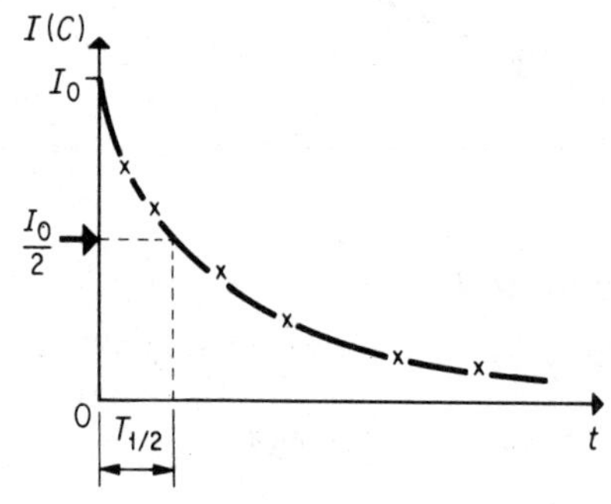

(*b*) Results

FIG. 18.7 Half-life and measurement.

Discovery of nucleus [Geiger and Marsden (1911)]. *Apparatus.* Source of α-particles (radon in tube), radiation limited to narrow beam, thin film of gold foil in evacuated vessel, microscope with zinc sulphide screen in front.

Observation. Most of the α-particles passed straight through the foil or were only slightly deflected; but some were scattered through large angles such as 30° or even 150°, as shown by scintillations on screen.

Conclusion. Rutherford concluded that some α-particles were violently repelled by a concentrated positive charge at the heart or centre of the atom, occupying a tiny volume, i.e. there was an atomic *nucleus* carrying positive electricity.

Rutherford calculated, on this basis, the number of α-particles which would be scattered in a particular direction. The formula was confirmed in subsequent experiments by Geiger and Marsden.

Discovery of proton in nucleus [Rutherford (1919)]. *Apparatus.* Vessel containing nitrogen, source of α-particles, silver foil sufficiently thick to absorb α-particles, fluorescent screen viewed by microscope.

Observations. Scintillations seen on screen.

Conclusions. Long range particles, more penetrating than α-particles, were ejected from nitrogen atoms. Comparison with range of protons emitted when α-particles incident on hydrogen atoms, and deflection of long range particles in magnetic field, showed *protons ejected from nitrogen atoms.*

Experiments also carried out with other elements, boron, fluorine, aluminium for example, showed that protons are present in the nuclei of all atoms.

Neutron. Discovered by Chadwick in 1931 in experiments with beryllium, which emits neutrons when bombarded with α-particles. He applied the laws of conservation of momentum and energy (assuming perfectly elastic collision) to the impact of neutrons on hydrogen and nitrogen nuclei, and calculated the mass of the neutron. Since a neutron has no charge, it is not repelled by an atomic nucleus. (An α-particle is repelled since it carries a $+$ve charge, like the nucleus.) Neutrons are therefore very penetrating particles and may hence produce nuclear fission (p. 150).

Some definitions. (*i*) An *electron* is a particle about 1/2000 of the mass of a hydrogen atom which carries the smallest negative charge e, numerically $1{\cdot}6 \times 10^{-19}$ coulomb.

(*ii*) A *proton* ($^1_1$H) is a hydrogen nucleus—it has a mass 1 and a charge $+e$.

(*iii*) A *neutron* ($^1_0$n) is a particle which has a mass practically equal to that of the proton but carries no charge.

(*iv*) A *nucleon* is a particle in the nucleus of an atom.

(*v*) An *isotope* is an atom whose nucleus has the same number of protons as another atom but a different number of neutrons. The chemical nature of the two atoms is the same.

(*vi*) The *atomic number* (Z) of an atom is the number of protons in its nucleus.

(*vii*) The *relative atomic mass* (A) of an element is the average mass of all its atoms compared with $\frac{1}{12}$ of the mass of the carbon atom ^{12}C.

(*viii*) The *mass number* of an atom is the number of protons and neutrons in the nucleus.

Einstein's mass-energy relation. A change m in mass produces an amount of energy E given by $E = mc^2$, where c is numerically equal to the velocity of light.

Units: E in joules (J) when m in kg and $c = 3 \times 10^8$ (m s^{-1})

Energy units. 1 electron-volt (eV) = work done or energy released when 1 electron moves through a p.d. of 1 volt = $1{\cdot}6 \times 10^{-19}$J.

1 a.m.u. = 1 atomic mass unit = 1/12th of mass of carbon atom, ^{12}C

$$= \frac{1}{12} \times \frac{12}{6 \times 10^{26}} \text{ kg} = \frac{1}{12} \times \frac{12}{6 \times 10^{26}} \times (3 \times 10^8)^2 \text{ J}$$

$$= 931 \text{ MeV}.$$

Binding energy. The protons and neutrons (nucleons) are kept together in the nucleus by the binding energy. This is equal to the energy required to just separate all the nucleons. From Einstein's law, the binding energy = difference in the total mass of individual nucleons and the mass of the nucleus.

Examples. (1) *Helium nucleus*, $^4_2$He. Mass of proton (p) = 1·0076 a.m.u., mass of neutron (n) = 1·009 a.m.u.

$$\therefore \quad \text{total mass of nucleons} = 2p + 2n = 4\cdot0332 \text{ a.m.u.}$$

But   mass of helium nucleus = 4·0028 a.m.u.

$$\therefore \quad \text{binding energy} = 4\cdot0332 - 4\cdot0028 \text{ a.m.u.}$$
$$= 0\cdot0304 \text{ a.m.u.} = 0\cdot0304 \times 931$$
$$= 28\cdot4 \text{ MeV}$$
$$\therefore \quad \text{binding energy per nucleon} = \frac{28\cdot4}{4} = 7\cdot1 \text{ MeV}.$$

(2) *Lithium disintegration* (Cockcroft and Walton). Protons ($^1_1$H) were accelerated by hundreds of thousands of volts to bombard lithium ($^7_3$Li). Two alpha-particles ($^4_2$He) were produced. $^7_3$Li + $^1_1$H $\rightarrow$ 2 $^4_2$He + Q, where Q is the energy released. The atomic masses of lithium and hydrogen = 7·018 + 1·008 = 8·026 a.m.u. The atomic mass of the α-particles = 2 × 4·004 = 8·008 a.m.u. Hence Q = 8·026 − 8·008 = 0·018 × 931 MeV = 16·8 MeV. Each α-particle had thus an initial energy of 8·4 MeV, as confirmed by measurements on the range of the α-particles.

Fission. A heavy nucleus such as $^{235}_{92}$U, uranium, breaks up into two relatively-heavy nuclei (fission) when struck by a neutron. Thus:

$$^{235}_{92}\text{U} + {}^1_0\text{n} \rightarrow {}^{148}_{57}\text{La} + {}^{85}_{35}\text{Br} + 3{}^1_0\text{n} + Q \quad . \quad . \quad . \quad . \quad (i)$$

Q = difference between masses on left side and right side. Calculation shows Q = 186 MeV = energy released per atom of uranium undergoing fission. In 235 kg of uranium, if all atoms undergo fission, energy released = 186 × (6 × 10^{26}) MeV, since there are 6 × 10^{26} (Avogadro constant) atoms in 1 kmol, 235 kg, of uranium.

Chain reaction. Several neutrons are produced in the fission reaction (see equation (i)). If their speed is moderated, the neutrons can be "captured" by other uranium atoms and produce further fission. The rapid multiplying fission effects, spreading through a mass of uranium, is a "chain reaction".

Fusion. In contrast to heavy elements, light elements can *fuse* together to form heavier elements and produce energy. Example: ^2_1H (deuterium) $+ {}^3_1H$ (tritium) $\rightarrow {}^4_2He$ (helium) $+ {}^1_0n + Q$. Fusion occurs if the deuterium and tritium, isotopes of hydrogen, can be heated to millions of degrees C. This is called a *thermonuclear reaction*. Thermonuclear reactions occur in the sun. Its energy is considered to be due to the fusion of hydrogen to form helium by a cycle of changes.

Biological effects. Natural chemical reactions of the body are upset by ionization of water, protein and other body fluids in chromosomes and cells. Tissue cells can be destroyed by γ-rays and energetic β-particles. *Precautions* include lead shielding against X- or γ-rays, containers of radioactive material with suitable shields, long tweezers for handling, and rubber gloves in handling solutions. Glass or perspex shields may be used in connection with β-particles.

Radiation units. 1. The *curie* (c) is the unit of radioactivity; it is equal to an activity of $3 \cdot 7 \times 10^{10}$ disintegrations per second.

2. The *röntgen* is a unit concerned with X- or γ-rays; it is the quantity of that radiation which produces $2 \cdot 08 \times 10^{15}$ ion pairs per m^3 of air under standard conditions of temperature and pressure.

3. The *rad* is a unit of radiation dosage concerned with α- or β-particles; it is the dose of radiation from these particles which produces absorption of $0 \cdot 01$ J per kg in any medium.

4. The *rem* is the dose of any radiation which produces the same biological effect as 1 röntgen of X- or γ-radiation.

$$Rems = Rads \times R.B.E,$$

where R.B.E. ("radiation biological efficiency") is a factor which takes into account the biological effect of the radiation.

19. X-RAYS. PHOTOELECTRICITY

Nature of X-rays. Unlike cathode-rays, which are particles, X-rays are *waves*. They are electromagnetic waves, like light waves, but of very short wavelength of the order 10^{-10} m. X-rays are thus not deflected by a magnetic or electric field. When they are passed through thin crystals or metal films, a diffraction (Laue) pattern emerges, proving that X-rays have wave properties.

X-ray tube. This consists of (*i*) a tungsten filament emitting electrons, (*ii*) a metal cylinder, the cathode, in which the filament is situated and which focuses the electrons on to the target, (*iii*) a target of tungsten or other metal in the end of a copper rod, the anode, (*iv*) a cooling arrangement for the anode (Fig. 19.1a).

Electrical supplies. A high voltage V, of the order of tens of thousands of volts, is applied between anode and cathode. The higher the voltage, the greater is the penetrating power of the X-rays; the "quality" of the rays is thus controlled by this voltage, which is obtained by means of a transformer and rectifying unit.

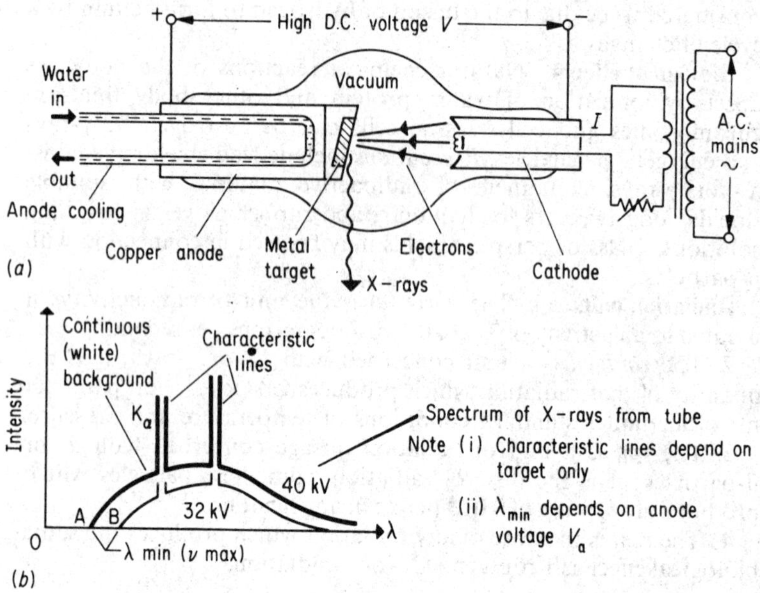

FIG. 19.1 X-ray tube. Characteristic spectra.

A low voltage, of the order of a few volts, is obtained from a separate small transformer, and applied across the filament. The higher the filament current I, the greater is the number of electrons per second emitted. The current thus controls the "intensity" of the X-ray beam, and is altered by means of a resistance in series with the secondary winding.

The *energy* of the arriving electrons $= \frac{1}{2}mv^2 = eV$, where v is the velocity and V is the anode potential, assuming the initial velocity of the electrons is zero. Here m is in kg, v in m s^{-1}, e in C, V in V.

X-ray characteristic spectra. These are the intense K or L lines, characteristic of the *metal target* of the X-ray tube. Fig. 19.1*b*. They are obtained when very energetic electrons penetrate deeply into the metal atoms and eject an electron from the innermost K shell for example. Fig. 19.2*a*. An electron from the L shell then falls into the K shell, thereby reducing the energy of the atom. The excess energy W is emitted in the form of radiation such that $W = h\nu$.

152

Since W is the energy change for electrons close to the nucleus, v depends on the nuclear charge, Ze, or on the atomic number Z. Moseley found *by experiments* with different target elements in an X-ray tube that, for the same K or L line, the graph of $v^{1/2}v$. Z was a straight line. Moseley's law is $v^{1/2} = a(Z-b)$, where a, b are constants. Fig. 19.2b.

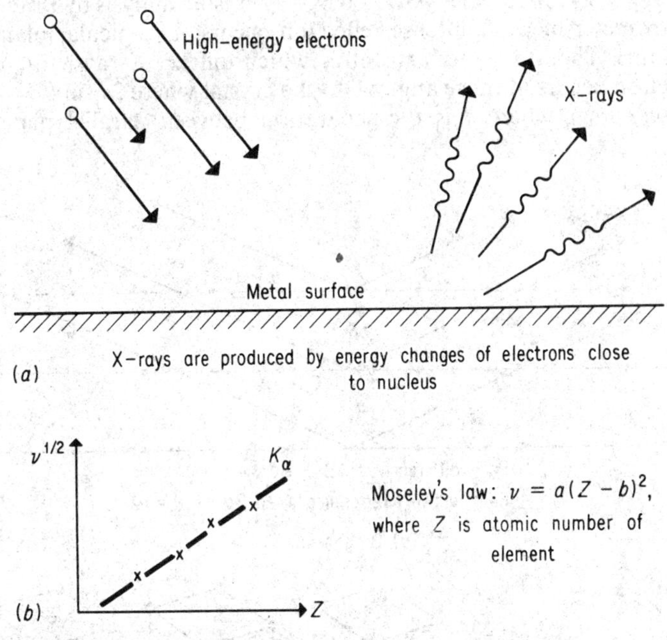

FIG. 19.2 X-ray and energy changes. Moseley's law.

X-ray continuous background.   In addition to intense K or L lines, X-ray spectra contain continuous ("white") radiation. Their wavelengths or frequencies depend on the p.d. V across the X-ray tube—they are independent of the nature of the metal target. *Explanation*: Electrons which strike the metal atoms or nuclei lose energy by the collisions, which may be glancing or direct collisions. On the quantum theory, any loss of energy W is emitted as radiation of frequency v given by $W = hv$. The greatest energy loss by atomic collisions is eV. The X-ray spectrum thus has a maximum frequency v_{max}, or a minimum wavelength λ_{min}, given by

$$W = eV = hv_{max} = \frac{hc}{\lambda_{min}}$$

$$\therefore \quad \lambda_{min} = \frac{hc}{eV}$$

153

For a voltage of 40 000 V, $e = 1.6 \times 10^{-19}$ C, $h = 6.6 \times 10^{-34}$ J s, $c = 3.0 \times 10^8$ m s^{-1}, then substituting in the above, we have

$$\lambda_{min} = \frac{6.6 \times 10^{-34} \times 3 \times 10^8}{40\,000 \times 1.6 \times 10^{-19}}$$

$$= 0.3 \times 10^{-10} \text{ m.}$$

Bragg's law. X-ray analysis. In X-ray crystal analysis by Bragg's spectrometer method, intense reflection occurs at particular planes of atoms. The atoms act as centres which diffract X-rays. Intense reflection occurs at those angles θ to the crystal where $2d \sin \theta = n\lambda$ (*Bragg's law*), where d is the separation between parallel planes.

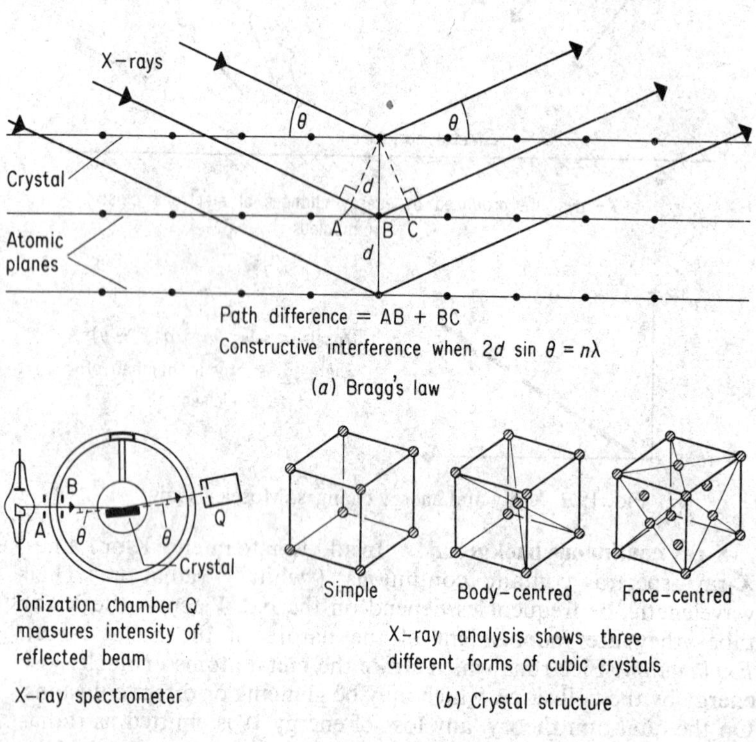

FIG. 19.3 Bragg's law and X-ray analysis.

Fig. 19.3a. The value of d can be found from the molecular weight, crystal density, and Avogadro constant (6×10^{26} per kmol), which together give the interatomic spacing a. If d is known, λ can be calculated by measuring θ. The known wavelength can then be used to find d for other crystals. Fig. 19.3b.

EXAMPLE

Potassium chloride has a relative molecular mass (molecular weight) of 74·5 and a density of 2000 kg m^{-3}. Hence number of molecules per metre^3 = (2000 kg/74·5 kg) $\times 6 \times 10^{26}$.

Each molecule has two atoms. Hence number of atoms in 1 m^3 = $2 \times 6 \times 10^{26} \times 2000/74·5$.

$$\therefore \quad \text{volume occupied per atom} = \frac{74·5}{2 \times 6 \times 10^{26} \times 2000} \text{ m}^3$$

$$\therefore \quad \text{average separation of atoms, } a = \left(\frac{74·5}{2 \times 6 \times 10^{26} \times 2000}\right)^{1/3}$$

$$= 3·1 \times 10^{-10} \text{ m}.$$

Wave nature of particles. De Broglie's law.   Slow-moving electrons produce a diffraction pattern on passing through a very thin metal film (Sir G. P. Thomson). TELTRON diffraction tube shows the ring pattern when electrons are diffracted through a carbon film.

Electrons (and other particles) have waves associated with them whose wavelength λ is given by

$$\lambda = \frac{h}{mv} \quad \text{(De Broglie's law)}$$

The quantity "mv" represents the momentum of the particle. If electrons, charge e, are accelerated to a velocity v by a p.d. V, then $\frac{1}{2}mv^2 = eV$, or $\lambda = h/\sqrt{2emV}$. Calculation shows that λ is several thousand times *shorter* than the wavelength of visible light when V is of the order of 5 kV say. Thus electron microscopes have resolving powers far greater than that of an optical microscope.

PHOTOELECTRICITY

Electrons are ejected from an illuminated metal plate. The experimental laws are: (1) The velocity of the electrons is independent of the intensity of the incident light. (2) Above a certain wavelength (threshold value) no electrons are emitted, even though the intensity is increased enormously.

Einstein's photoelectric theory.   This states that light consists of particles (photons) whose energies E are proportional to the frequency v of the light, or $E = hv$, where h is Planck's constant. Fig. 19.4.

(1) The energy to just liberate electrons from surface atoms of a metal is known as the *work function*, w_0, of the metal, and is a constant for a particular metal.

(2) Thus $hv - w_0 =$ maximum energy of liberated electron, $\frac{1}{2}mv_m^2$

(3) The threshold value, v_0, is given by $hv_0 = w_0$, or $v_0 = w_0/h$.

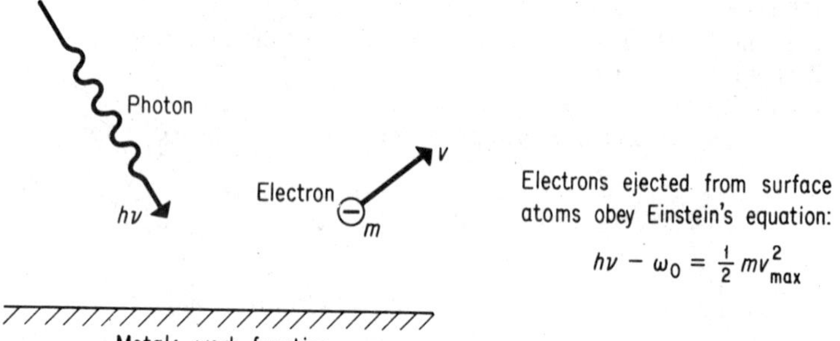

FIG. 19.4 Photoelectric effect and theory.

Millikan's experiment and deduction. Sodium, potassium and lithium emit electrons when illuminated by visible light. The metal surfaces such as A were cleaned by K, and illuminated by monochromatic light of frequency v. Fig. 19.5a. The stopping potential,

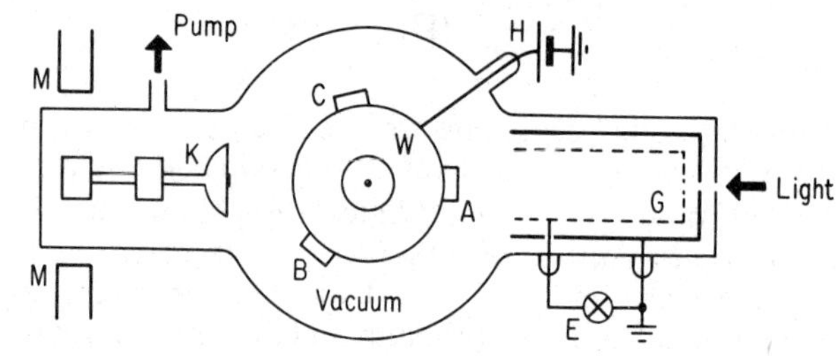

(a) Apparatus

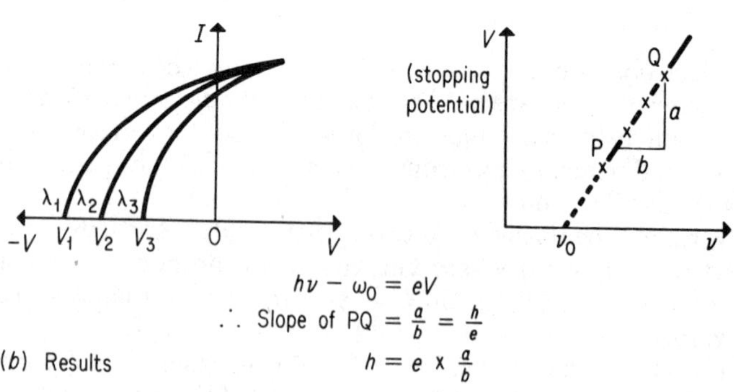

(b) Results

$$hv - w_0 = eV$$
$$\therefore \text{ Slope of PQ} = \frac{a}{b} = \frac{h}{e}$$
$$h = e \times \frac{a}{b}$$

FIG. 19.5 Millikan's photoelectric experiment.

156

V, which just prevented electrons from being collected by G, was measured. Now if v_m is the maximum velocity of the electrons,

$$hv - w_0 = \tfrac{1}{2}mv_m{}^2 = eV.$$

When V was plotted against v a straight line was obtained. Fig. 19.5b. The gradient of the line was h/e, and knowing e, then h was calculated. Its value was found to be close to 6.6×10^{-34} J s, or close to Planck's constant known from experiments on heat radiation.

Duality of wave and particle. Light may therefore be considered as waves *or* particles (photons). Mathematically, a wave can be made equivalent to a particle, and either concept can be used as required. Electrons have wavelike properties—diffraction rings are produced when incident on very thin gold films (G.P. Thomson). They are similar to X-rays in this respect. De Broglie suggested that the wavelength λ of moving particles is given by h/mv, where h is Planck's constant and mv is the momentum of the particle. Other particles than electrons, such as molecules, have wavelike properties. Sound waves can be treated, if required, as particles (photons).

Photoelectric cell. Fig. 19.6a illustrates a vacuum-type of *photoelectric cell*; electrons emitted from C are attracted to A since A has a positive potential relative to C. For a given wavelength of incident light, the current is proportional to the intensity of the light.

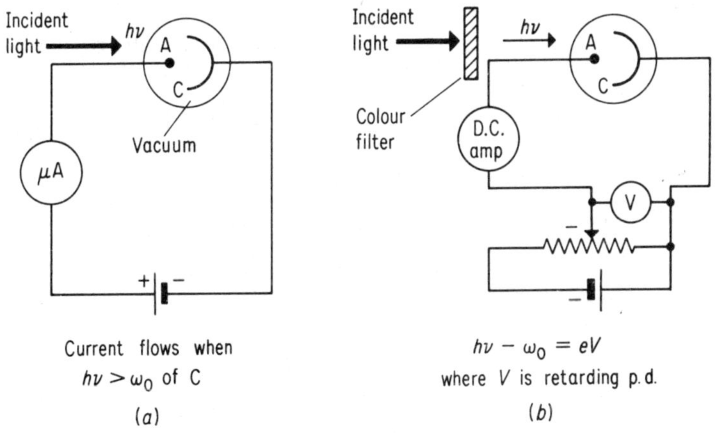

FIG. 19.6 (*a*) Photoelectric cell. (*b*) Measurement of h.

Fig. 19.6b illustrates the principle of an experiment to measure h using a photoelectric cell. This time the potential of A can be increased *negatively* relative to A until the current is zero. If V is the "stopping potential" then $eV = hv - w_0$. The frequency v of the incident light can be varied by using pure colour filters.

157

20. ENERGY LEVELS IN THE ATOM

Franck-Hertz experiment. *Object.* To bombard atoms of gases such as neon, helium, sodium vapour by electrons of known energy, and to see if the energy of an atom can be raised in definite amounts, i.e. to investigate if the energy in the atom is "quantized".

Experiment. The Franck-Hertz tube has gas at a very low pressure (1 mmHg); an accelerating p.d. V between C, G; a small *retarding* p.d., less than 1V, between G and the collecting anode A; a current meter for measuring anode current I. Fig. 20.1a.

(a) Apparatus

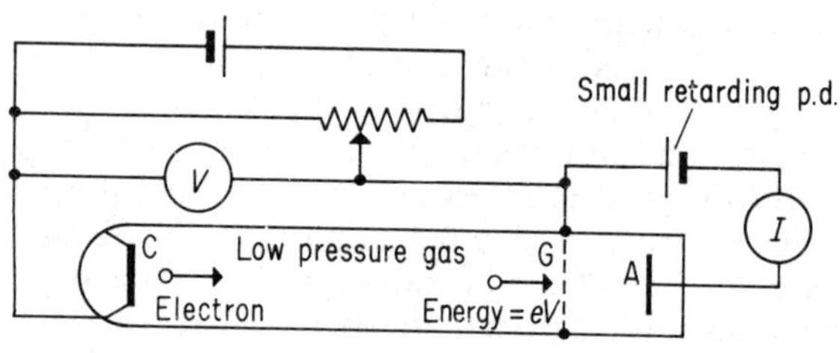

(b) Results

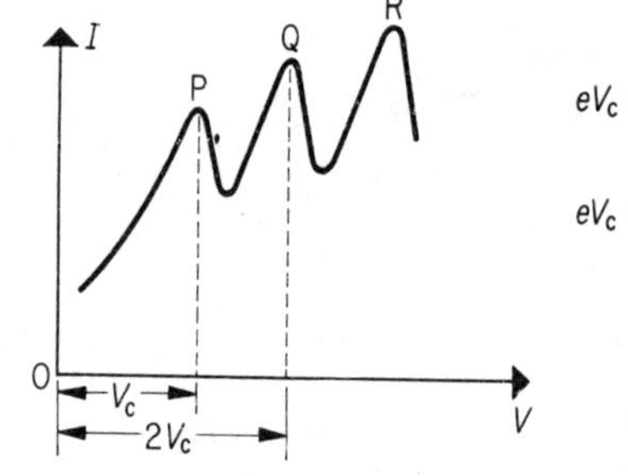

eV_c = energy when <u>inelastic</u> collisions first occur and current falls at P

$eV_c = E_1 - E_0$ = difference in energy levels of atom = $h\nu$, where ν is emitted photon frequency

(c) Explanation

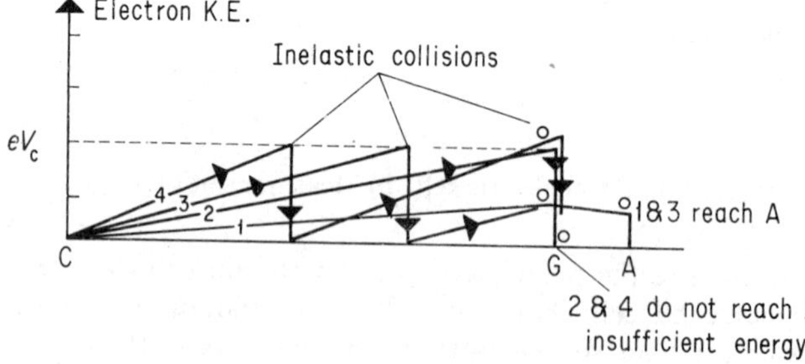

FIG. 20.1 Accelerated electrons. Elastic and inelastic collisions.

Results. A series of peaks and dips in the anode current *I*v.*V* curve as *V* is increased from zero, showing that some electrons have lost energy and are unable to reach A at certain values of *V*. Fig. 20.1*b*.

Conclusions. (1) Inelastic collisions occur between many of the bombarding electrons and atoms at definite energies of the electron, i.e. energy is lost by the electron and given to the atom, so that the atom's energy is raised. Fig. 20.1*c*. (2) The drop in current at the peaks occurs at definite changes V_c of p.d. Hence the energy of the atom can only be raised in definite amounts. The "energy levels" of the atom are thus quantized; they are separated levels and the atom cannot have energies between these values. Fig. 20.2*a* shows energy levels, E_0, E_1, E_2, . . ., of the hydrogen atom.

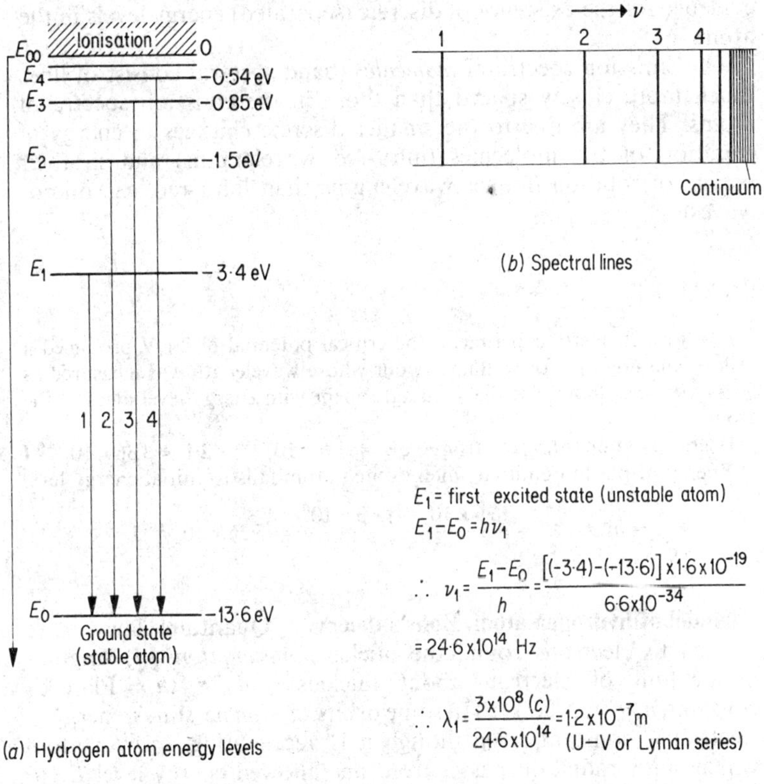

$$E_1 = \text{first excited state (unstable atom)}$$
$$E_1 - E_0 = h\nu_1$$
$$\therefore \nu_1 = \frac{E_1 - E_0}{h} = \frac{[(-3\cdot4)-(-13\cdot6)]\times1\cdot6\times10^{-19}}{6\cdot6\times10^{-34}}$$
$$= 24\cdot6\times10^{14}\ \text{Hz}$$
$$\therefore \lambda_1 = \frac{3\times10^8\ (c)}{24\cdot6\times10^{14}} = 1\cdot2\times10^{-7}\text{m} \quad \text{(U−V or Lyman series)}$$

FIG. 20.2 Energy levels and radiation.

Excitation, ionization potentials. *Ground state* of atom = lowest energy state or level = most stable energy level. Energy given to an atom may bring it to a higher energy level or *excited state*. If the atom absorbs an amount of energy eV_i sufficient to just remove the most loosely-bound electron, V_i is the first *ionization potential* of the atom.

Exitation and ionization potentials are called *critical potentials.* If V_c is the critical potential of the gas atom in a Franck-Hertz tube, the energy of an electron making an inelastic collision with it is eV_c.

Spectra. If the energy of the atom is raised to a level E_n above that of its ground state E_0, the atom is unstable in this excited state. When it falls to a lower energy level, say E_0, the loss of energy is radiated as a photon of energy $h\nu$, where $\nu = c/\lambda =$ the frequency of electromagnetic radiation, i.e. $E_n - E_0 = h\nu$. Fig. 20.2b.

When the atom falls to the ground state, a series of discrete spectral lines are emitted. Each line has a frequency proportional to the energy change from one level to another. Thus optical spectra is evidence for the existence of discrete (separated) energy levels in the atom.

The emission spectra of *molecules* (band spectra) consist of lines much more closely spaced than those in the emission spectra of atoms. They are due to the smaller discrete changes in energy of vibration of the molecules (infra-red wavelengths) and in their energy of rotation (longer wavelengths than infra-red, e.g. micro-waves).

EXAMPLE

In a Franck-Hertz experiment, the critical potential of 2·1 V produced a yellow emission line for sodium vapour whose wavelength was measured as $5\cdot89 \times 10^{-7}$ m. Show that this is in agreement with energy level ideas of the atom.

Energy of bombarding electron $= eV = (1\cdot6 \times 10^{-19}) \times 2\cdot1 = 3\cdot36 \times 10^{-19}$ J

Energy of photon emitted when excited atom falls to initial energy level

$$= h\nu = \frac{hc}{\lambda} = \frac{(6\cdot6 \times 10^{-34}) \times 3 \times 10^8}{5\cdot89 \times 10^{-7}} = 3\cdot36 \times 10^{-19} \text{ J}$$

Model of hydrogen atom. Bohr's theory. Quantum laws were applied to electrons round the nucleus. *Assumptions.* (*i*) Angular momentum of electron about nucleus $= nh/2\pi$ ($h =$ Planck's constant, $n = 1, 2, \ldots$). (*ii*) In some orbits, *stationary states,* energy of electron is constant, even though it is accelerating. (*iii*) An atom which emits radiation passes from one allowed energy level E_1 to another E_2 of smaller value, the difference in energy, $E_1 - E_2$, being emitted in the form of electromagnetic radiation of frequency ν, where $E_1 - E_2 = h\nu$.

Calculation. $mv^2/r = mr\omega^2 = e^2/4\pi\varepsilon_0 r^2$ (force equation), $mvr = nh/2\pi$ (angular momentum equation), $E_1 - E_2 = h\nu$ (energy and radiation equation). Fig. 20.3.

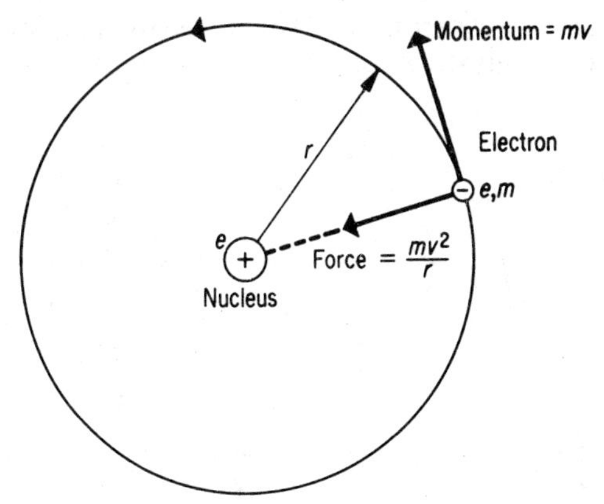

Force = $\dfrac{mv^2}{r}$ = $\dfrac{e^2}{4\pi\varepsilon_0 r^2}$

Angular momentum = mvr = $\dfrac{nh}{2\pi}$

Energy $E = \frac{1}{2}mv^2 - \dfrac{e^2}{4\pi\varepsilon_0 r}$

$E_1 - E_2 = h\nu$

Deductions

1. Orbits quantized

2. Spectral series given by

$$\bar{\nu} = \frac{\nu}{c} = R\left(\frac{1}{n_2{}^2} - \frac{1}{n_1{}^2}\right)$$

3. $r = 0.53 \times 10^{-10}$ m, frequency of rotation $\simeq 10^{15}$ rev s^{-1}

FIG. 20.3 Bohr's theory of hydrogen atom.

On calculation,

$$r = \frac{\varepsilon_0 n^2 h^2}{\pi m e^2} \qquad \ldots \ldots \ldots \quad (1)$$

$$\bar{\nu} = \text{wave number}\left(\frac{1}{\lambda}\right) = \frac{m e^4}{8\varepsilon_0{}^2 c h^3}\left(\frac{1}{n_2{}^2} - \frac{1}{n_1{}^2}\right) \quad \ldots \quad (2)$$

Deductions. (*a*) When $n = 1$ (most stable orbit), $r = \varepsilon_0 h^2/me^2 = 5.3 \times 10^{-11}$ m (approx.) on substituting values of m, e, h. This is in good agreement with r found by kinetic theory of gases. (*b*) *Rydberg's constant*, R, $= me^4/8\varepsilon_0{}^2 ch^3$. On calculation, excellent agreement with observed value. (*c*) *Spectra of hydrogen.* Lyman (U–V), Balmer (visible), Paschen (I–R) series all agree with formula (2).

Electron shells. Electrons round the nucleus of an atom occupy one of a number of closely-spaced discrete energy levels called *shells*. Pauli's Exclusion Principle showed that up to 2 electrons occupy the K shell, nearest to the nucleus; the next, L, shell can be occupied up to a maximum of 8 electrons; the next, M, up to a maximum of 8 electrons; and so on. The total number of electrons = the atomic number, Z, of the atom. Fig. 20.4.

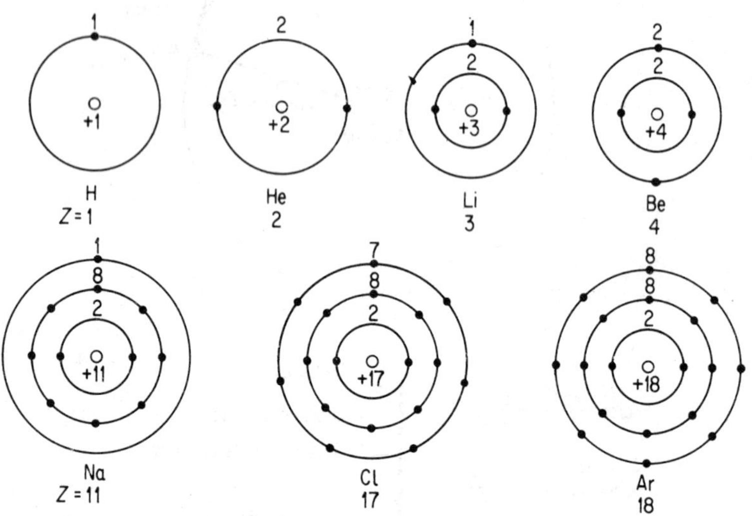

Shell	K	L	M	N	O	P
Sub-shell no	2	2,6	2,6	2,6,10	2,6,10	2,6,10,14
Element	He	Ne	Ar	Kr	Xe	Rn
Z	2	10	18	36	54	86

FIG. 20.4 Electron shells in atoms.

Chemical activity. Sodium ($Z = 11$) has 2 electrons in the K shell (full), 8 electrons in the L shell (full), 1 electron in the M shell. Sodium is therefore an electron "donor". Chlorine ($Z = 17$) has 2, 8, 7 electrons in the K, L, M shells respectively. Chlorine is therefore an electron "acceptor". Sodium chloride is hence a stable compound; the outermost electron is transferred between the two atoms. Lithium and potassium are chemically similar to sodium— they have 1 electron in their outermost shell. Fluorine and bromine are chemically similar to chlorine— they have a vacancy for 1 electron in their outermost shell. Neon ($Z = 10$) has 2, 8 electrons in the K, L shells respectively, both of which are therefore full. Hence neon is a chemically inert gas. If a graph is plotted of the first ionization potentials of elements against their atomic number, alkali metals such as lithium, sodium all have minimum values (outermost electrons form a new subshell and are loosely bound).

The inert gases such as helium and neon all have peak values of ionization potential (all their subshells are complete and the outermost electrons are strongly bound). Fig. 20.5.

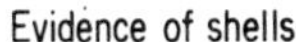

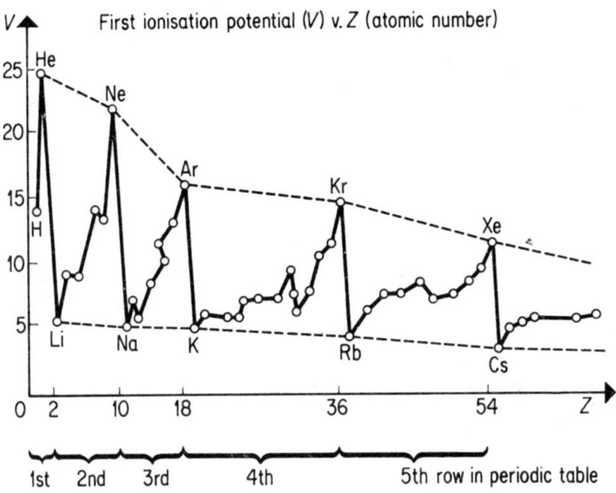

Peak values indicate complete electron sub-shells (rare gases)
Minimum values indicate loosely bound outermost electron (alkali metals) and beginning of new sub-shells, or rows in periodic table.

Note: Alkali metals are conductors, rare gases are insulators

FIG. 20.5 Ionisation potential and chemical activity.

Optical spectra are due to energy level changes of electrons in the outermost shells, as already stated. *X-ray spectra* are due to electrons ejected from the innermost shells, K or L, to other shells. When the atom returns to its ground state, high-frequency or X-rays are emitted as the energy change is high.